TECHNICAL SKETCHING
WITH AN
INTRODUCTION
TO CAD

TECHNICAL SKETCHING WITH AN INTRODUCTION TO CAD

For Engineers, Technologists, and Technicians

Third Edition

DALE H. BESTERFIELD, *PH.D., P.E.*
Professor Emeritus, College of Engineering
Southern Illinois University

ROBERT E. O'HAGAN, *M.S. IND. TECH.*
Assistant Professor Emeritus, College of Engineering
Southern Illinois University

Prentice Hall
Upper Saddle River, New Jersey   Columbus, Ohio

Library of Congress Cataloging-in-Publication Data

Besterfield, Dale H.
    Technical sketching for engineers, technologists, and technicians
/ Dale H. Besterfield, Robert E. O'Hagan. — 3rd ed.
      p.   cm.
    Includes index.
    ISBN 0-13-472572-7
    1. Freehand technical sketching.   I. O'Hagan, Robert E.
II. Title.
T359.B47   1997
604.2—DC21

97–14321
CIP

Editor: Stephen Helba
Production Editor: Rex Davidson
Production Service: The Clarinda Company
Design Coordinator: Julia Zonneveld Van Hook
Cover Designer: Proof Positive/Farrolyne Associates
Production Manager: Laura Messerly
Marketing Manager: Debbie Yarnell

This book was set in Souvenir by The Clarinda Company and was printed and bound by Courier/Kendallville, Inc. The cover was printed by Phoenix Color Corp.

© 1998, 1990, 1983 by Prentice-Hall, Inc.
Simon & Schuster/A Viacom Company
Upper Saddle River, New Jersey 07458

Printed in the United States of America

10 9 8 7 6 5 4 3 2 1

ISBN: 0-13-472572-7

Prentice-Hall International (UK) Limited, *London*
Prentice-Hall of Australia Pty. Limited, *Sydney*
Prentice-Hall Canada, Inc., *Toronto*
Prentice-Hall Hispanoamericana, S. A., *Mexico*
Prentice-Hall of India Private Limited, *New Delhi*
Prentice-Hall of Japan, Inc., *Tokyo*
Simon & Schuster Asia Pte. Ltd., *Singapore*
Editoria Prentice-Hall do Brasil, Ltda., *Rio de Janeiro*

CONTENTS

PREFACE

There has been a significant change in the method of communicating technical information in graphical form. A substantial number of engineers, technologists, and technicians use sketches to communicate their ideas. Computer-aided drafting and design has taken the place of instrument drawings, and the T-square, triangle, and compass are as obsolete as the slide rule.

Even though the computer has replaced instrument drawings, technical sketching is still used to generate initial ideas on paper. A knowledge of technical sketching is a prerequisite to learning computer-aided drafting and design and is essential to interpreting working drawings generated by a computer.

This book has been designed to enable the reader to:

1. Communicate, by means of a three-view or pictorial sketch, an idea or concept. The sketch should be sufficiently detailed, clear, and accurate to enable working drawings to be produced by a computer.
2. Read and interpret a working drawing.
3. Understand the basic principles of graphic communication as an initial step toward acquiring proficiency in computer-aided drafting and design.

Technical programs in community colleges and technical institutes, such as machine tool, automotive, electronics and electrical, maintenance, and welding, to name a few, will find this book an excellent one to meet their needs. Most engineering technology and industrial technology programs in colleges and universities will find that the sketching approach will also meet their needs. The book is ideal for a freshman course or part thereof in an engineering program. It will also serve the instructional needs of in-house programs in industry either on a group or self-instruction basis.

The book begins with chapters on sketching techniques and lettering. This is followed by a detailed description of orthographic projection, pictorials, auxiliary views, and sectioning. Additional chapters cover dimensioning, tolerancing, fastening techniques, and working drawings. The final chapters cover CAD, and there is a new chapter on Autocad. All the essential areas that are covered in a textbook on instrument drawing are included. Sufficient exercises are given to enable the reader to become proficient in the technique.

The third edition has been improved by updating the dimensioning and tolerancing material to conform to the new standards. All chapters have been updated. Metric units are used throughout. Additional exercises have been included where appropriate.

The authors are grateful to the numerous students and faculty throughout the world who have helped to eradicate errors and clarify explanations.

We are also grateful to Glen Besterfield, who improved Chapter 11, and to Ron Rathje for his contribution of Chapter 12.

Dale H. Besterfield
Fairview Hts., IL

Robert E. O'Hagan
Fond du Lac, WI

SKETCHING

CHAPTER

1

Upon completion of this chapter, the student is expected to:

- Know the purpose of sketching.

- Know the suggestions for good technical sketching procedures and habits.

- Be able to construct an acceptable free-hand sketch.

INTRODUCTION

The importance of technical sketching in engineering and technology cannot be overemphasized. Many original design ideas are conceived by means of sketches. Engineers frequently turn sketches of their ideas over to drafters to construct computer-generated drawings or to technicians to construct the actual object.

Technical sketching is used as a means of rapid graphical communication that enables the engineer to quickly express, explain, and record her or his ideas. Sketching is done freehand. All sketches are made with dark, distinct lines and are suitable for reproduction by photocopy or ozalid reproduction.

The principal difference between a technical sketch and a technical drawing is one of degree. Sketches are made freehand, whereas drawings are made by the computer.

SKETCHING MATERIALS

One advantage of technical sketching is that it only requires pencil and paper. The types of pencils are grouped into hard, medium, and soft leads. Hard leads make a light thin line, soft leads make a dark wide line, and medium leads produce a line between these two extremes. Numerals and letters are used to designate the type of lead. An HB-grade lead (medium) is recommended to sketch lines that describe the shape of an object and also to letter. A 3H-grade lead (hard) is used to sketch lines that aid in the construction of the object. The medium-grade leads are sufficient for most sketching applications. Lead grades in the medium category are 3H, 2H, H, F, HB, and B, with the B grade being softer than the H grades. The texture of the paper and the humidity may require a slightly different grade of lead. Experimentation is recommended to find the grade that provides the best results. Types of lines are discussed at the appropriate point in the text.

The pencil should be kept sharp at all times. Mechanical pencils with thin leads are available that eliminate the need to sharpen the lead.

Due to recent advances in fiber pens, practitioners have found it convenient to construct an

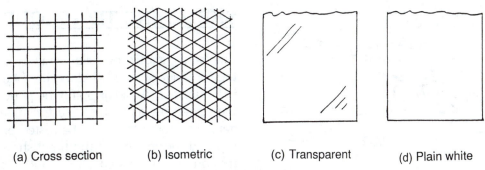

(a) Cross section   (b) Isometric   (c) Transparent   (d) Plain white

FIGURE 1-1
Types of paper.

initial sketch with penciled lines, and then to produce dark distinct lines using the appropriate fiber pen. Many designers use various templates to construct circles, ellipses, and various symbols of fluid power, welding, electrical system parts, and so forth.

The other sketching material is paper. Essentially, there are four types of sketching paper: cross section, isometric, transparent, and plain white. They are illustrated in Figure 1-1.

Cross-section or grid paper is the most common sketching paper. Horizontal and vertical lines can be easily sketched by following the grid lines. Different grid spaces, such as four, five, eight, and ten squares to the inch, are available. Special cross-section paper using blue grid lines that do not reproduce by photocopy or other reproduction methods can be obtained. This special paper has the advantage of the grid lines for sketching and a reproduced copy that does not have the grid lines.

Isometric paper is used for a particular type of pictorial sketch. The three different grid-line directions correspond to the three axes of the isometric pictorial. This type of paper is also available with grid lines that do not reproduce.

Transparent paper, which is sometimes referred to as onion skin, tracing, or vellum, has the advantage that light passes through it. Sketch pads of transparent paper are available with various types of master grids. When the master grid is inserted beneath the transparent paper, the grid lines are visible. The sketch is constructed using these grid lines; however, the finished sketch does not show the grid lines. Plain white paper can be used for sketching; however, the other types are easier to use.

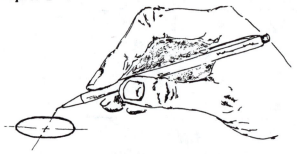

FIGURE 1-2
Holding the pencil.

SKETCHING TECHNIQUES

The pencil should be held firmly in the hand with the forefinger extended along the pencil as shown in Figure 1-2. Avoid gripping the pencil tightly with the fingertips, as this is very tiresome on the fingers. The forearm should be rested on a flat tabletop and the motion of the hand should be a wrist motion when making short- and medium-length lines. Avoid sketching positions that do not rest the hand and arm on the tabletop. Grip the pencil at a normal length from the end, approximately as shown in the illustration. It is not necessary to fasten the paper to the table.

The technique for sketching horizontal lines is shown in Figure 1-3(a). It requires a wrist motion for medium or short lines, not a whole-arm motion. Lines should be drawn from left to right for right-handed people and right to left for left-handed people.

It is considered best to draw vertical lines with a downward motion, as shown in Figure 1-3(b). This depends somewhat on the ability and preference of the individual person. Many people find it convenient to turn the paper to a position that is comfortable. The basic rule for sketching is to pull the pencil rather than push it against the paper.

One of the best techniques for sketching a line between two points is to start with the pencil at one point and draw the line quickly to the second point, keeping your eye on the second point, as

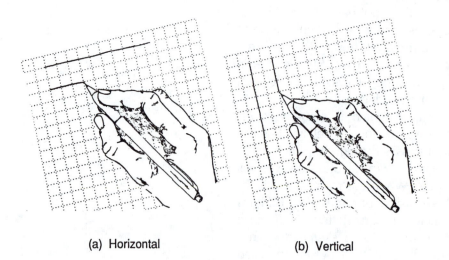

(a) Horizontal

(b) Vertical

FIGURE 1-3
Sketching lines.

shown in Figure 1-4. It is just as easy to make a solid continuous line as to make a series of short broken lines. Practice this technique and you will achieve excellent results.

Straight lines are the most common type of geometric construction. The circle is the second most common, and the two types account for approximately 95% of the geometric constructions. The recommended practice for sketching circles is shown in Figure 1-5 and consists of four steps:

Step I. Draw the lines that indicate the center of the circle.

Step II. Lay out the radii on each of these lines, thereby locating four points on the circumference of the circle.

Step III. Locate approximately two points in each quadrant equal to the radius of the circle.

Step IV. Draw in the solid line representing the circumference of the circle.

Many practitioners have found it more convenient to use a circle template, which is faster and neater than the foregoing method.

A three-viewed sketch of an object, using cross-section paper as an aid, is illustrated in Figure 1-6. Notice that the sketch lines stand out above the grid lines. Since the sketch is sometimes used in place of a drawing, it must be complete in every detail, including notes and dimensions.

Three types of lines are shown in Figure 1-6.

Visible or object lines describe the visible surfaces in the various views of the object. They

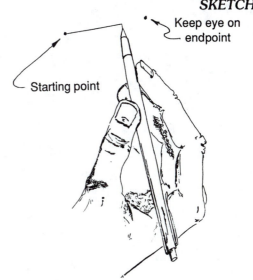

FIGURE 1-4
Line between two points.

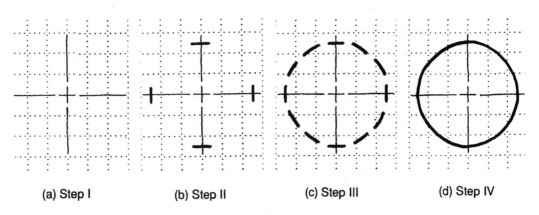

    (a) Step I   (b) Step II   (c) Step III   (d) Step IV

FIGURE 1-5
Method of sketching circles.

are thick dark lines and are constructed with an HB-grade lead. Visible lines are usually the widest and most prominent lines of a sketch.

Hidden lines describe the hidden surfaces in the various views of an object. They are dark thin lines constructed by alternating a short dash [about 3 mm (0.12 inch) long] and a space [1 mm (0.04 inch)]. An HB-grade lead can be used provided that the line is thin. If this is not possible, an H grade may be more appropriate.

Centerlines locate the center of a circle or an axis of a cylinder. They are constructed by alternating a short dash [about 3 mm (0.12 inch) long], a space [about 1 mm (0.04 inch)], and a long dash between 20 mm (0.80 inch) and 40 mm (1.58 inches) long. The center of the circle or of an arc is located by the crossing of two short dashes. An HB-grade lead can be used provided that the line is thin. If this is not possible, an H grade may be more appropriate. Centerlines are shown in all views of an object.

It is not necessary to be an artist to make an acceptable sketch. With practice, sketched lines can be constructed with almost the same appearance as lines drawn with the computer.

PROPORTION

Although a sketch is not made to a precise scale, it must be in the proper proportion. An object may be sketched actual or full size; however, if the ob-

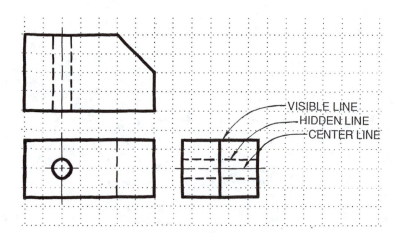

FIGURE 1-6
Three-view sketch of an object.

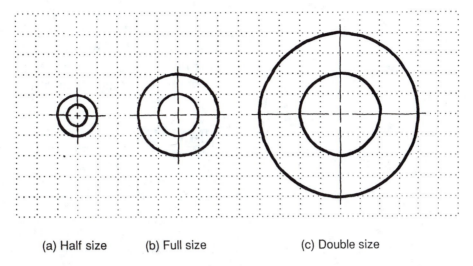

(a) Half size   (b) Full size   (c) Double size

FIGURE 1-7
Proportion.

ject is quite large, it will need to be sketched smaller. If the object is quite small, it will need to be sketched much larger than its actual size. The proportion used depends on the relative size of the object and the size of the paper on which it is sketched. Figure 1-7 illustrates a half-size, an actual- or full-size, and a double-size sketch of a steel washer. Other proportions can also be used.

In sketching an object such as an automobile or some irregularly shaped part, it is often a decided help to use the grid system. First the main controlling points or proportionate size of the object is established on the grid. In this case, Figure

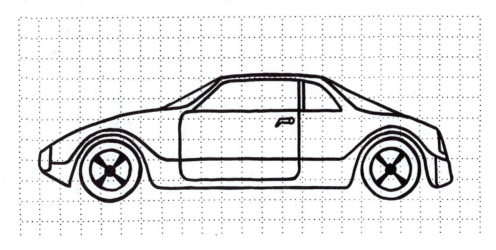

FIGURE 1-8
Use of grid lines to obtain proportion.

1-8, the bumper-to-bumper length of the automobile, the wheel base, and the height are established. Then the detail parts are sketched. The use of the grid system enables us to keep the parts of the automobile in their proper proportion.

SUMMARY

In summary, let's review the information that has been presented.

1. Use an HB-grade lead or fiber pen for sketching the shape of the object.

2. Sketches quickly convey, explain, and record ideas.

3. Technical sketches are not done to a precise scale; however, they are made in the proper proportion.

4. Lines on sketches should be dark and distinct and of the correct type.

5. Paper is not fastened to the table while sketching.

6. When possible, sketch lines should be made in one continuous stroke.

7. Visible lines represent visible surfaces; hidden lines represent hidden surfaces; and centerlines locate the center of a circle or an axis of a cylinder.

8. Cross-section paper and isometric paper are designed to aid in sketching.

9. Sketches must be complete in every detail, as they are sometimes used in place of computer drawings.

10. Sketches should be of such quality that they can be reproduced.

11. The pencil should be pulled rather than pushed.

12. It is not necessary to be an artist to make an acceptable sketch; however, practice will develop improved proficiency.

LETTERING

CHAPTER

2

Upon completion of this chapter, the student is expected to:

• Be able to convey, on a sketch, verbal and numerical information that is legible, neat, and well proportioned.

INTRODUCTION

This unit deals with the fundamentals and techniques of good lettering. It is a requirement on all sketches that the lettering be legible, neat, and well proportioned. The proper pencil must be used in lettering, and the character and proportions of the letters must be within reasonable standards. This chapter emphasizes vertical lettering, guidelines, and fractions. Inclined lettering and lowercase lettering are not included.

Guidelines are necessary to achieve uniform lettering. The square grid of cross-section paper can be used for guidelines. Lettering is accomplished using an HB-grade lead. With a reasonable amount of conscientious practice, good lettering can be achieved.

VERTICAL LETTER ALPHABET

Both capital letters and numerals for vertical-style lettering are shown in Figure 2-1. The reader should study the proportion of the various letters—width to the height. We can see from this illustration that the letter **I** is the narrowest, while **W** is the widest. A good approach is to separate the letters and numerals into their common groups: the straight-line group and the curved-line group. The two strokes to master are the straight-line stroke and the oval stroke. Once this is accomplished, we are able to form most of the letters either as single strokes or as combinations of straight-line and oval strokes.

The order of strokes used to form the various straight-line letters is shown in successive steps in Figure 2-2. The letter is shown at the left, and successively to the right are the steps taken in forming this letter. It is well to analyze these one by one for the proper execution of the letter by the orderly

ABCDEFGHIJKLMNOPQRS
TUVWXYZ 0123456789

FIGURE 2-1
Vertical letter alphabet.

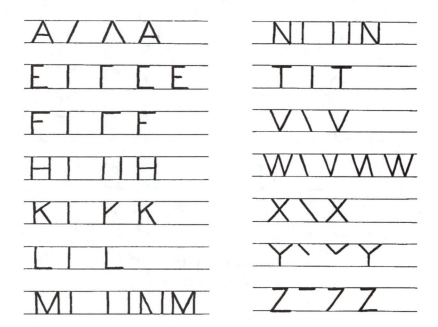

FIGURE 2-2
Order of straight-line strokes.

succession of strokes. Strokes are made from top to bottom and left to right for right-handed persons and right to left for left-handed persons so that the pencil is pulled rather than pushed.

The oval-group letters and the successive steps in forming the letters are shown in Figure 2-3.

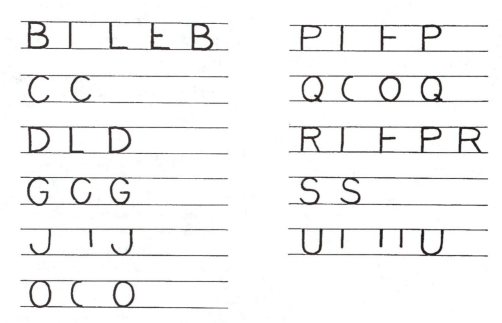

FIGURE 2-3
Curved-line lettering strokes.

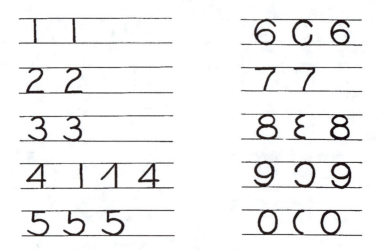

FIGURE 2-4
Numerical strokes.

Each letter should be analyzed for the character of the letter, the proportions, and orderly step-by-step execution of forming the letter.

The numerals, as shown in Figure 2-4, should be analyzed one by one and the proper order of strokes studied and practiced. The height of the numerals is the same as that of the letters. Where fractions are shown, the whole number of the fraction is the same height as the letters.

Stability or pleasing proportion of lettering is an important part of the letter or numeral. The top portion of **3, 8,** and **B** should be a little smaller than the bottom to give the appearance of stability.

The need to follow the proper order of strokes cannot be overemphasized.

GUIDELINES

Guidelines are a must to obtain legible, neat, and well-proportioned letters and numerals. For sketching, the square grid of cross-section paper is usually used for the guidelines, thereby eliminating the need for their construction. A square grid of 5 mm (0.20 inches) is usually satisfactory and is shown in Figure 2-5.

If a square grid is not available, guidelines must be constructed using a straightedge and 6H-grade lead.

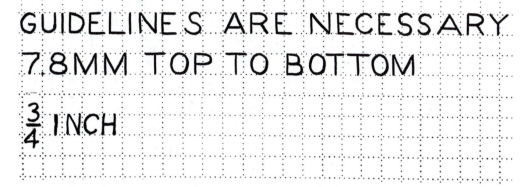

FIGURE 2-5
Use of the square grid for guidelines.

SPACING

To obtain a pleasing appearance, the spacing between words and letters is critical. Spacing between letters is a function of the area between the letters, not the distance between the letters. The spacing between words should be approximately the width of the capital letter **O,** as shown in Figure 2-6. The correct (upper line) and incorrect (lower line) spacing between letters is also shown. To make the letters and the space between letters exceptionally wide is poor practice in that they take up too much space. To make the letters and the space between letters too narrow makes them too hard to read and they tend to blur in reproduction. Attempts should be made to keep the ar-

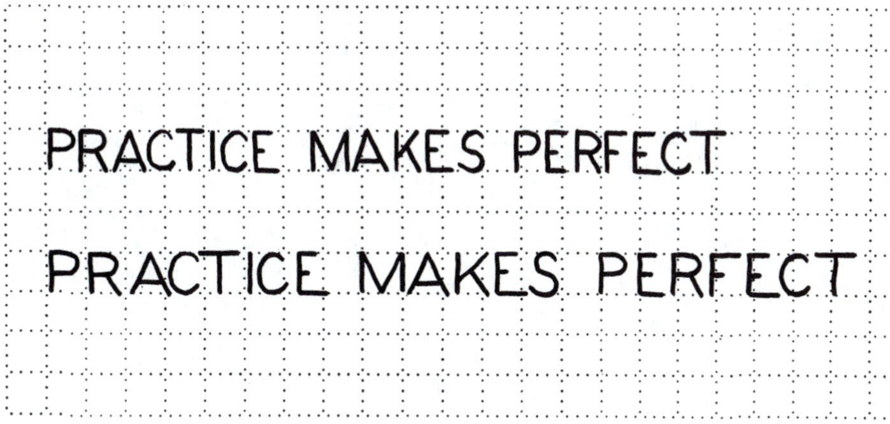

FIGURE 2-6
Lettering examples.

eas between letters and the width of letters uniform and approximately as shown in the figure. The space between lines should be equal to the height of letters.

SUMMARY

In summary, one must know or do the following:

1. Know the character and proportions of the letter or numeral to be made.

2. If a square grid is not available, use a 6H-grade lead for guidelines. Use an HB-grade lead for lettering.

3. Use proper spacing between letters and proper spacing between words.

4. Follow the recommended order of strokes for forming the letter or numeral.

5. Strokes are made top to bottom and left to right for right-handed people and right to left for left-handed people.

6. Practice sufficiently to acquire reasonable proficiency.

ORTHOGRAPHIC PROJECTION

CHAPTER

3

Upon completion of this chapter, the student is expected to:

- Know the principles of orthographic projection.

- Know the rules for selecting the front view.

- Be able to identify the different types of surfaces.

- Be able to sketch three views of an object from a pictorial sketch.

- Be able to sketch a principal view of an object, given the other two views.

INTRODUCTION

Orthographic projection is a graphical means of communication used by engineers, technologists, and technicians. Special training is necessary in the use of this means of communication. It is a graphical language, and because of the basic principles used in developing the language, it is interpreted and understood by people of different nationalities with different written and spoken languages.

An understanding of orthographic projection involves a study of the basic rules for developing the views of an object, coupled with analysis, visualization, and reasoning. Neither making nor interpreting the orthographic views of an object is a mechanical process.

Orthographic projection is used to create, design, describe, manufacture, fabricate and assemble, and repair and replace parts, machines, and concepts. Its origin dates back to Biblical times, and future use is assured. Our industrial, technical, and scientific society is built on the ability to communicate by orthographic projection. Orthographic projection is also referred to as multiview drawings or sketches. The orthographic concept is used in sketching and computer graphics.

TYPES OF PROJECTIONS

Sketches of the different types of projections are shown in Figure 3-1. All these projections are of the same object. The most commonly used projection is the orthographic.

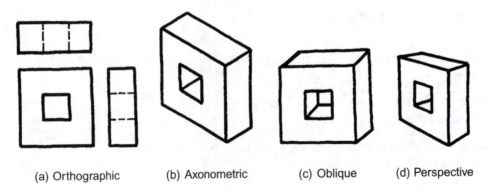

(a) Orthographic   (b) Axonometric   (c) Oblique   (d) Perspective

FIGURE 3-1
Types of projections.

The other three projections represent the object in three-dimensional form. **Axonometric** projection has a definite relationship of the three axes and the length along each axis. There are three types of axonometric projection. The most common one is called isometric and it is discussed in Chapter 4.

Oblique projection is actually not a true projection. For some objects it will convey the same information as an axonometric projection and is easier to construct. Oblique sketches are also discussed in Chapter 4.

The **perspective** type of projection is similar to a photograph; however, it is difficult to construct and sketch. Its principal use is in architecture.

PRINCIPLE OF ORTHOGRAPHIC PROJECTION

The fundamental principle of **orthographic** projection involves the projection of an image of the object on an imaginary plane by use of projectors that are perpendicular to that plane and parallel to each other. The image of the object is called the orthographic view. For this to be theoretically possible, the observer is considered to be at an infinite location. The plane is called a projection plane. This concept is illustrated in Figure 3-2.

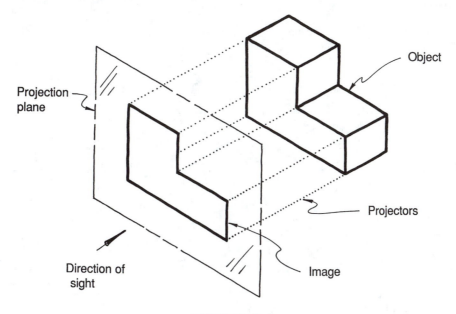

FIGURE 3-2
Projection of an object.

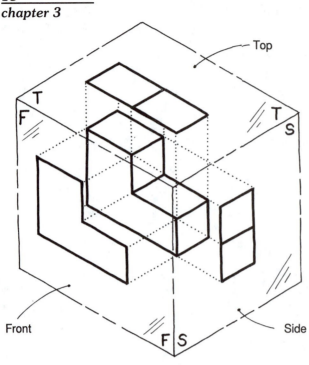

FIGURE 3-3
Projection box.

One of the best ways to understand the concept of orthographic projection is by means of an imaginary glass box or projection box, as shown in Figure 3-3. The object is inside the projection box and the image of the object is projected to the sides of the box, which are the projection planes. Visible projection planes are the front, top, and side. The six sides of this imaginary glass box are the projection planes, and they are at right angles to each other. Orthographic projection means projection in which the projection lines are perpendicular (at right angles) to the projection plane. _Ortho_ means at right angles, and _graphic_ means describing in full detail.

There are two systems of projection that are recognized internationally: first-angle projection and third-angle projection. In third-angle projection the object is behind the projection planes as described in the preceding two paragraphs. Third-angle projection is the system used in the United States and Canada. In first-angle projection the object is in front of the projection planes, as shown in Figure 3-4. It is the system used in the rest of the world.

Since it is required to show views of a solid or three-dimensional object on a flat sheet of paper, it is necessary to unfold the six projection planes so that they all lie in a two-dimensional plane such as this page. The six sides of the glass box form the

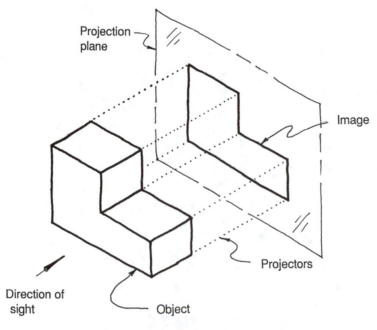

FIGURE 3-4
First-angle projection.

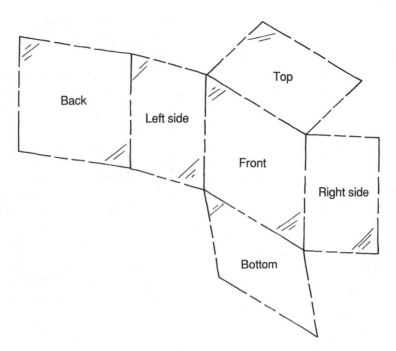

FIGURE 3-5
The six primary planes of projection.

six primary planes of projection. Front, top, right-side, bottom, left-side, and back projection planes are shown in their unfolded state in Figure 3-5.

The image of the object is projected to the planes as shown in Figure 3-6. While there are six primary views, pairs of views are the same. The

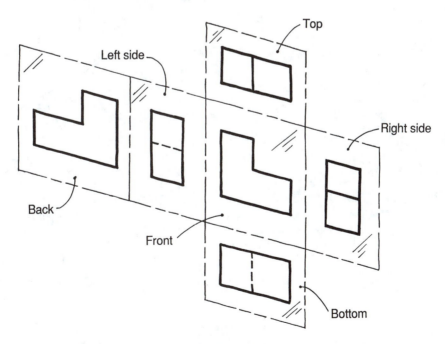

FIGURE 3-6
The six primary planes of projection and the object.

front and back are the same, except that they are mirror images. The top and bottom views are the same, except that in the top view the vertical surface in the center is represented by a visible line and the bottom view by a hidden line. The right-side and left-side views are the same, except that in the right-side view the horizontal surface of the object is represented by a visible line, and in the left-side view by a hidden line.

Since pairs of views are the same except for visible and hidden surfaces, three projections will usually describe the object. These projections are the front, top, and right side and are called the standard views. In the remainder of the book the right-side view will be referred to as the side view. The left side view would be used if it would require fewer hidden lines. The front view is the principal view.

Our discussion of the concept continues with a different object enclosed by the imaginary glass box. Figure 3-7 shows an object with normal, cylindrical, and inclined surfaces. The hole is shown, using hidden or dashed lines in two of the views. Centerlines are used to describe the location of the center of the hole and the axis of the cylinder. The projection planes with the image of the object are partially unfolded, as shown in the figure. Only three views are needed to describe the object.

Note the use of folding lines that correspond to the hinge lines of the glass projection box. Between

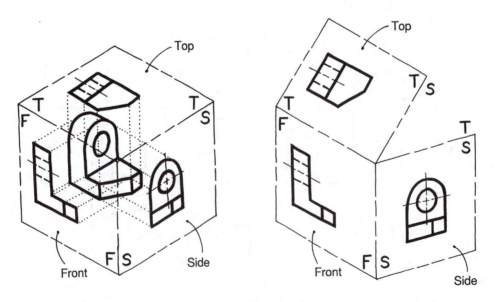

FIGURE 3-7
Projection box and three planes unfolded.

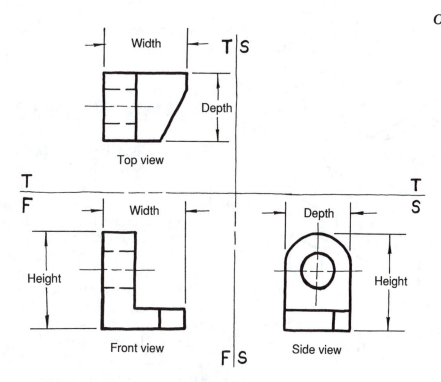

FIGURE 3-8
Orthographic or multiview sketch.

the top projection plane and the front projection plane is the folding line labeled T/F. Between the front projection plane and the side projection plane is the folding line labeled F/S. Between the top projection plane and the side projection plane is the folding line labeled T/S. In the unfolded position the T/S folding line has two positions.

These folding lines are the edges of the projection planes. For example, the folding line T/F is the edge view of the front projection plane when viewed from the top and the edge view of the top projection plane when viewed from the front. In each view, two of the projection planes appear as an edge.

When the projection planes are unfolded 90°, the result is three views on the two-dimensional page as shown in Figure 3-8. The folding lines are reference lines, which are used in the construction of the three-view sketch. They can be erased after the sketch is completed.

There is a definite relationship between the views shown by the lightly drawn lines with the arrows. The width of the object is shown in the front and top views, the height in the front and side views, and the depth in the top and side views. Each feature of the object, such as a hole, is shown in each view. Also each feature is the correct dis-

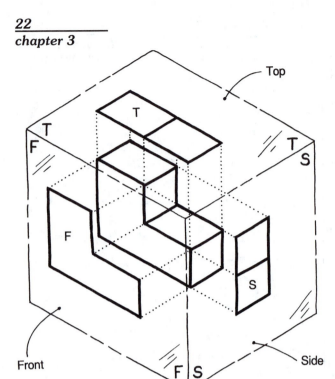

FIGURE 3-9
Projection box and object.

tance from the projection planes as represented by the folding lines.

VIEWING AN OBJECT

The projection box and our original object are shown in Figure 3-9. Also shown are the orthographic views on the three faces of the projection box. We will now develop the principles of projection for each view. If you have understood the previous material, this information will be a review. If you are still a little hazy, this section will clarify your thinking on the principle of orthographic projection.

A pictorial sketch and a three-view orthographic sketch are illustrated in Figure 3-10. The image of the object is projected to the front projection plane by level lines of sight in space. This image, called the front view, is sketched by the observer, who is assumed to be at an infinite distance. The top and side views are shown to indicate their relative position in the three-view sketch.

The observer now moves to the top of the projection box to obtain the top view as shown in Figure 3-11. The image of the object is projected to the top projection plane by vertical lines of sight in space. This image, called the top view, is sketched by the observer, who is assumed to be at an infinite distance. The front and side views are also shown to indicate their relative position in the three-view sketch.

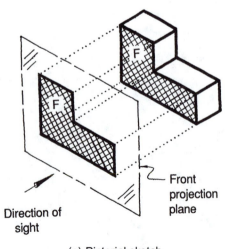

(a) Pictorial sketch

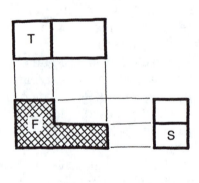

(b) Orthographic views

FIGURE 3-10
Front view.

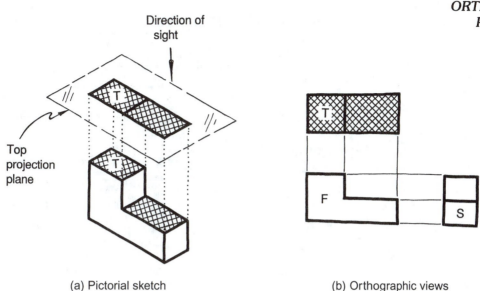

(a) Pictorial sketch

(b) Orthographic views

FIGURE 3-11
Top view.

The observer moves to the side of the projection box as shown in Figure 3-12 to obtain the side view. The image of the object is projected to the side projection plane by level lines of sight in space. This image, called the side view, is sketched by the observer, who is assumed to be at an infinite distance. The front and top views are also shown to indicate their relative position in the three-view sketch.

It is important to remember that the object inside the projection box is left stationary. The object is never rotated; the observer moves to each position.

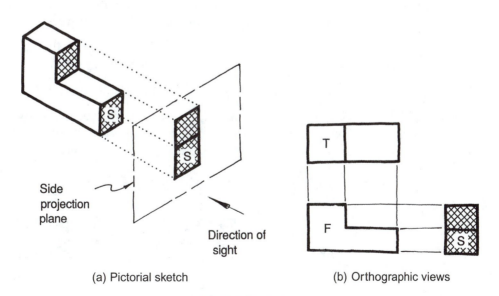

(a) Pictorial sketch

(b) Orthographic views

FIGURE 3-12
Side view.

REVIEW

At this point we can review the principles and basic rules for making orthographic sketches.

1. Orthographic projection means right-angle projection.

2. Orthographic projection uses projection planes.

3. Each set of two adjacent projection planes is perpendicular to each other.

4. The lines of sight (projectors) are parallel to each other and perpendicular to the projection plane.

5. The observer is assumed to be at an infinite distance from the projection plane.

6. A projection is the image of the object on the projection plane.

7. The three most commonly used projections are front view, top view, and side view.

8. The front view is the principal view.

9. Only one view of an object may be seen at one time by an observer at an infinite distance.

10. Three views are usually necessary to describe an object; however, some objects might only need two or one.

11. Two of the planes of projection appear as an edge in each orthographic view.

12. Folding lines represent the edges or intersections of projection planes.

13. The object is stationary; it is never rotated.

14. A three-view sketch means that the observer has viewed the object, at an infinite distance, from three different directions.

SELECTING THE BEST CHOICE OF THE FRONT VIEW

The person making the sketch selects the best choice for the front view. The object is placed in a stable position and turned so that the characteristic shape shows and unnecessary hidden surfaces are avoided. Figure 3-13(a) illustrates the best choice of the front view for the L-shaped object.

The best choice for the front view for a different object is shown in Figure 3-13(b). This front view shows the object to be rectangular in shape and has a circular hole running through the object. This is the proper choice for the front view of such an object.

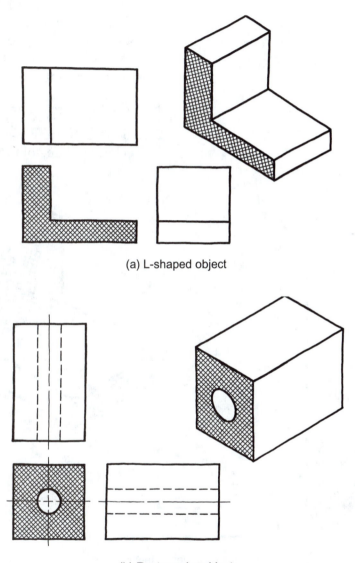

(a) L-shaped object

(b) Rectangular object

FIGURE 3-13
Best choice of front view.

A poor choice of selecting the front view is shown in Figure 3-14(a). The front view does not give the characteristic shape of the object; it is the side view that shows the L-shape of the object.

Figure 3-14(b) shows a different choice for the front view. It, too, is a poor choice. Although we have the characteristic shape shown in the front view, we have introduced an unnecessary hidden line in the side view.

Figure 3-14(c) also shows a poor choice of a front view. The object is in an unstable position.

The selection of the front view is very important since the proper choice makes the three-view sketch easier to read. Rules for selecting the front view are as follows:

1. The front projection is the principal view.

2. Choose the front view to show the most characteristic shape of the object.

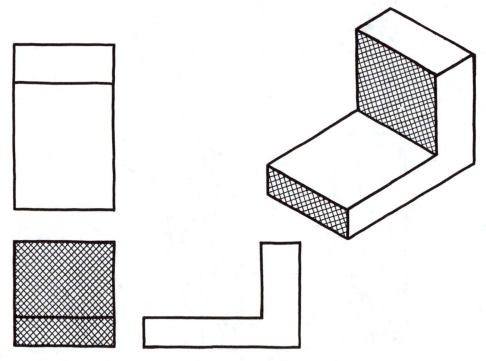

(a) Does show characteristic shape

FIGURE 3-14
Poor choice of front view.

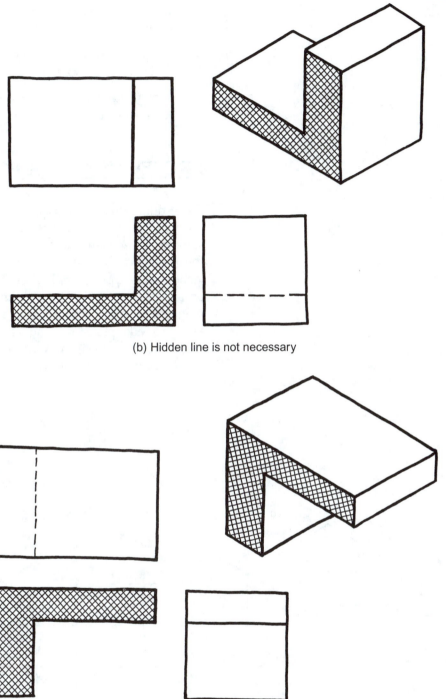

(b) Hidden line is not necessary

(c) Object is not stable

FIGURE 3-14, Continued

3. The object should be sketched in a stable position.

4. The object should be sketched to see as many visible lines as possible in the front view.

5. If only two views are needed, the front view with either the top or side view is used.

PROJECTION OF A POINT

Since numerous points make up a line or surface, the projection of a point is fundamental to understanding three-view sketches. A point, 1, is inside the projection box as shown in Figure 3-15. Projections of the point are labeled 1 in the front, top, and side views. The folding line represents the plane as an edge. The top plane (folding line T/F) and the side plane (folding line F/S) appear as edges when the object is viewed from the front. Likewise, when the object is viewed from the top,

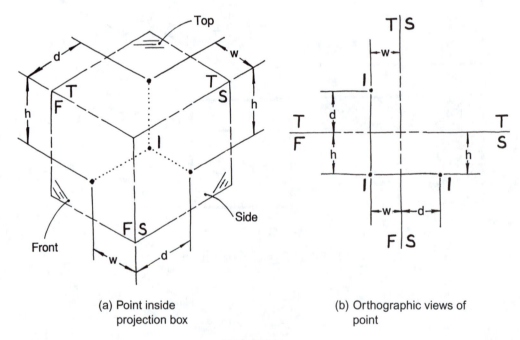

(a) Point inside projection box

(b) Orthographic views of point

FIGURE 3-15
Projection of a point.

the side plane (folding line T/S) and the front plane (folding line T/F) appear as edges. Furthermore, when the object is viewed from the side, the top plane (folding line T/S) and the front plane (folding line F/S) appear as edges. Folding line T/S appears in both the top view and the side view. It is very important to realize that in any given view you are looking perpendicular to one plane and seeing the two adjacent planes as edges.

There are certain common distances in the related views. When looking down on point 1, it will appear at distance *d* (depth) back of the front plane (folding line T/F). This same distance back of the front plane (folding line F/S), distance *d* will show in the side view and may be located that way. Notice that the height of the point, distance *h* (height), is the same in the front and side views. In other words, they are on a level horizontal line, parallel to the T/F folding line, drawn across from the front to the side view. The distance point 1 is from the side plane is given by the letter *w* (width). Emphasis should be placed upon which planes appear as an edge in which of the views and where the distance can be measured in respect to that plane.

The folding lines actually become the separation lines for the views. They form a cross whereby the vertical line is the folding line T/S at the top and folding line F/S at the bottom. The horizontal line is the folding line T/F at the left and folding line T/S at the right.

The reader should study Figure 3-15 closely, as the projection of a point is the foundation of orthographic projection. Lines and surfaces are merely projections of points.

Usually, the point in the front view is located first, and then the top view distance *d*. The point in the top view is located a distance *d* from folding line T/F and is directly above and directly in line with the front view, as shown in Figure 3-16(a). The side view of the point is drawn last and is in a fixed relationship to the other views. There are three ways of locating the side projection. It may be accomplished with a 45° projection line, as shown in Figures 3-16(b) and 3-16(c). This is the most common method of projecting points to the side view. It may be accomplished by a radius, as shown in Figure 3-16(d). What we are doing is merely transferring the distance from the front projection plane as it appears in the top view to the same distance from the front projection plane as it appears in the

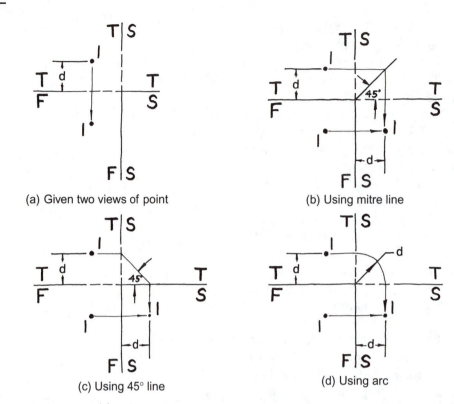

(a) Given two views of point   (b) Using mitre line

(c) Using 45° line   (d) Using arc

FIGURE 3-16
Locating the side projection.

side view, and this could be accomplished using just the grid lines. Remember that the other coordinate of the point in the side view is obtained by sketching a level horizontal line, parallel to the folding line. In complicated sketches, it is frequently necessary to label points to obtain the side view. This practice will be discussed later in the chapter.

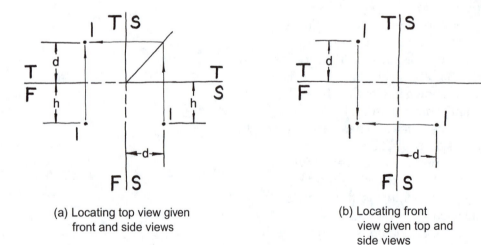

(a) Locating top view given front and side views

(b) Locating front view given top and side views

FIGURE 3-17
Locating the top and front projections.

If the point is located first in the front view and then in the side view, the same technique can be used as shown in Figure 3-17(a). The point in the side view is located a distance d from folding line F/S and the same distance h below the horizontal folding lines (T/F and T/S).

The simplest transfer occurs when a point is located in the top and side views and is transferred to the front view. This type of transfer is shown by the thin lines with the arrows in Figure 3-17(b).

PROJECTION OF A LINE

There are different types of lines, such as normal, inclined, oblique, and curved. The same principle of point projection can be extended to line projection.

A **normal line** is parallel to two of the projection planes and perpendicular to the other one. Assume a horizontal line in the glass projection box. The line in the front view as shown in Figure 3-18(a) is parallel to the folding line and is the correct length. Since the line is parallel to the front projection plane, the line is true length (TL) in the front view. The line in the top view must be directly above the front view. The line in the top view is also true length because it is parallel to the top projection plane. Once the front and top views are sketched, the projection to the third view is fixed.

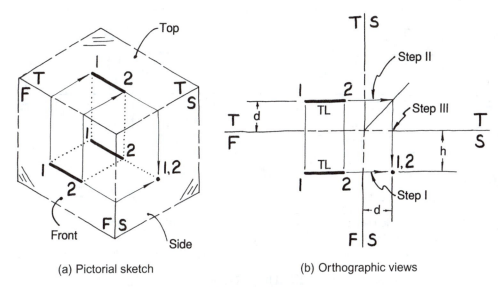

(a) Pictorial sketch   (b) Orthographic views

FIGURE 3-18
Projection of a normal line to the side view.

Step I is to project a horizontal line from the front view to the side view. Step II is to project a horizontal line from the top view to the 45° line, and Step III is to project down to the side view. The intersection of these two projections is the point labeled 1,2. In this case the line is perpendicular to the side projection plane and, therefore, appears as a point in the side view. Obviously, it is not a true-length projection in the side view. When a line is parallel to two of the three projection planes, it will appear true length in two of the views and as a point in the third.

In the preceding paragraph the normal line was projected from the front and top views to the side view. If the normal line is located first in the front view and then in the side view, the line is projected to the top view as illustrated in Figure 3-19(a). The normal line (appears as a point) in the side view is located a distance *d* from the folding line F/S and a distance *h* below the horizontal folding lines T/F and T/S. Projection to the top view is illustrated by the thin lines with arrows.

Similarly, if a normal line is given in the top view, it is located in the side view a distance *h* from folding line T/S and a distance *d* from folding line F/S. Projection to the front view is shown in Figure 3-19(b) by the thin lines with arrows.

A normal line can be parallel to the front plane and side plane as shown in Figure 3-20(a). Also, it can be parallel to the top plane and to the

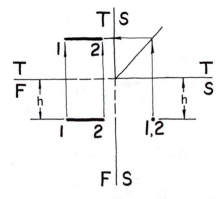

(a) Locating side view given
top and front views

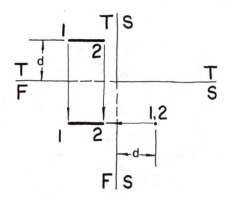

(b) Locating front
view given top and
side views

FIGURE 3-19
Projection of a normal line to the top
and front views.

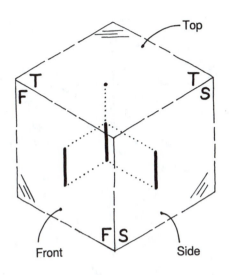

(a) TL line in front and
side views

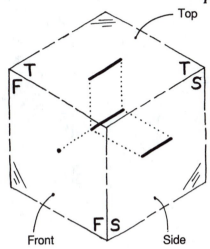

(b) TL line in top
and side views

FIGURE 3-20
Normal lines perpendicular to the top
and front planes.

side plane as shown in Figure 3-20(b). The projections of these lines are left as an exercise.

Now let's consider a line that is parallel to only one side of the glass projection box, say the front projection plane as shown in Figure 3-21(a). This line is called an **inclined line** and is parallel to only one projection plane. The line is sketched in the front view with the correct slope and length.

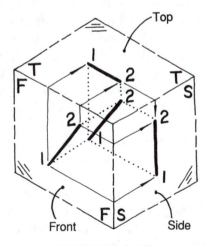

(a) Pictorial sketch

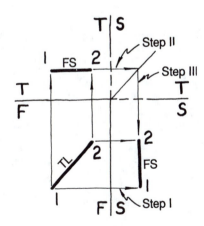

(b) Orthographic views

FIGURE 3-21
Projection of an inclined line to
the side view.

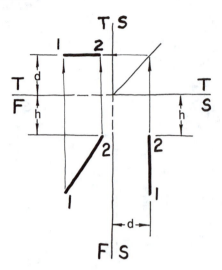

(a) Finding top view given front
and side view

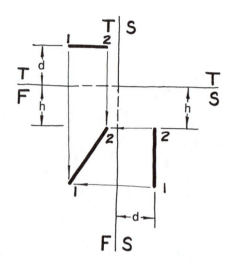

(b) Finding front view
given top and side view

FIGURE 3-22
Projection of an inclined line to the top
and front views.

The ends of the line are projected to the top view. Since the line is parallel to the front projection plane, it appears as a foreshortened (FS) line in the top view. The line is projected to the side view by steps I to III in the same manner as in Figure 3-18; however, in this case the line does not appear as a point but as a foreshortened line.

In the preceding paragraph the inclined line was projected from the front and top views to the side view. If the inclined line is located first in the front view and then in the side view, the line is projected to the top view as illustrated in Figure 3-22(a). The inclined line in the side view is located a distance d from the folding line F/S and a distance h below the horizontal folding lines (T/F and T/S). Projection to the top view is illustrated by the thin lines with arrows.

Similarly, if an inclined line is given in the top view, it is located in the side view a distance h from folding line T/S and a distance d from folding line F/S. Projection to the front view is shown in Figure 3-22(b) by the thin lines with arrows.

Lines that are parallel to only one of the three planes appear as true length in one of the views and foreshortened in the other two views. If the inclined line had been parallel to the top projection plane as shown in Figure 3-23(a), it would have appeared true length in the top view; if the line had been parallel to the side projection plane as shown

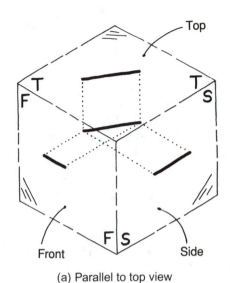

(a) Parallel to top view

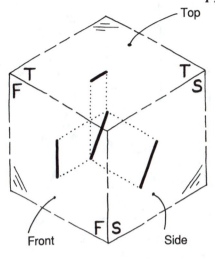

(b) Parallel to side view

FIGURE 3-23
Inclined lines parallel to the top
and side planes.

in Figure 3-23(b), it would have appeared true length in the side view. The three-view sketches of these two situations are left as exercises.

An **oblique line** is neither parallel nor perpendicular to any of the projection planes. The projection of an oblique line from the front and top views to the side view is shown in Figure 3-24. If the oblique line is located first in the front view, the ends of the line are projected to the top view. The location of ends 1 and 2 must be the correct distance from the front plane. Projection

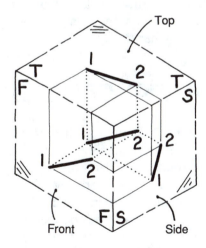

(a) Pictorial sketch

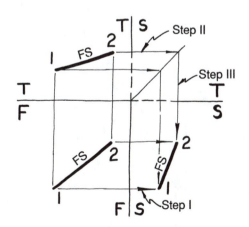

(b) Orthographic views

FIGURE 3-24
Projection of an oblique line
to the side view.

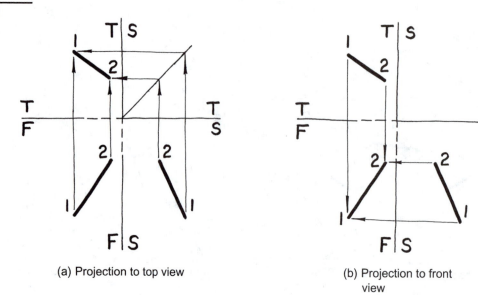

(a) Projection to top view

(b) Projection to front view

FIGURE 3-25
Projection of oblique lines to the top
and front views.

to the side view is shown by the thin lines with arrows.

Projection of an oblique line from the front and side views to the top view is shown in Figure 3-25(a) by the thin lines with arrows. Similarly, the projection of an oblique line from the top and side views to the front view is shown in Figure 3-25(b).

It is important for you to know that when a line is not parallel to any of the projection planes, the line is foreshortened in the front, top, and side views. To obtain the true length of this type of line, an auxiliary view is necessary. Auxiliary views are discussed in Chapter 6.

PROJECTION OF SURFACES

Although the study of projection of points and lines is important to understand the concept of orthographic projection, objects are actually constructed or manufactured by surfaces. When visualizing objects, the technically trained person thinks in terms of surfaces. An object can be made up of normal, inclined, oblique, and curved surfaces.

A **normal surface** is a plane surface of any shape that is parallel to a projection plane. The normal surface appears true size (TS) and shape in the plane to which it is parallel and a vertical or

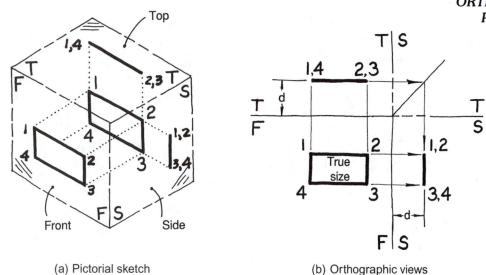

(a) Pictorial sketch   (b) Orthographic views

FIGURE 3-26
Normal surface projection.

horizontal line on the other two projection planes. Many normal surfaces are rectangles; however, the shape could be triangular, circular, octagonal, and so forth. The projection of the normal rectangular surface, as illustrated in Figure 3-26, is identical to the projection of four normal lines. First, sketch the rectangle in the front view (since it is parallel to that view). Next, the rectangle is located in the top view directly above the front view. It will appear as a line and can be located a distance d from the folding line. From there on the procedure is merely to project the four points of the rectangle to the side view and join them by lines. This particular normal surface is parallel to the front projection plane. Normal surfaces are also parallel to the side or top projection planes. Figure 3-27(a) shows a normal surface (shaped as a triangle) that is parallel to the top projection plane. Figure 3-27(b) shows a normal surface (shaped as a hexagon) that is parallel to the side projection plane. Sketching these surfaces in the three views is left as an exercise.

An **inclined surface** is a plane surface of any shape that is perpendicular to one projection plane but inclined to the others. Figure 3-28 has the inclined surface perpendicular to the front projection plane. The procedure for the projection of the surfaces is similar to the procedure for a normal surface. It relies on the concepts given previously in the projection of a point and line sections of this chapter.

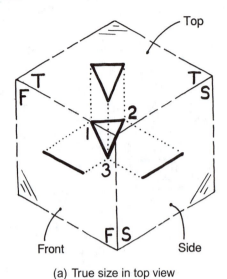

(a) True size in top view

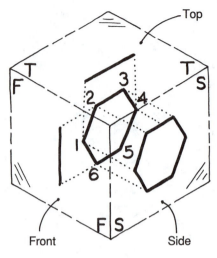

(b) True size in side view

FIGURE 3-27
Normal surfaces parallel to the top
and side planes.

This inclined surface is perpendicular to the front projection plane; an inclined surface can also be perpendicular to the side or top projection planes. Figure 3-29(a) shows an inclined surface (shaped as a triangle) that is perpendicular to the top projection plane. Figure 3-29(b) shows an inclined surface (shaped as a hexagon) that is perpendicular to the side projection plane. Sketching these surfaces in the three views is left as an exercise.

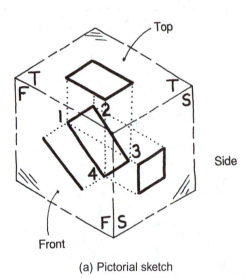

(a) Pictorial sketch

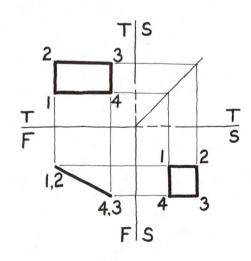

(b) Orthographic views

FIGURE 3-28
Inclined surface projection.

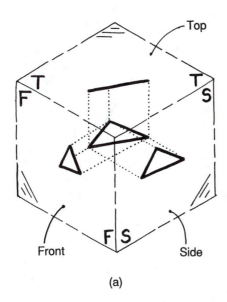

(a)

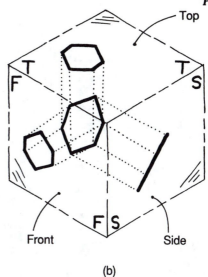

(b)

FIGURE 3-29
Inclined surfaces perpendicular to the
top and side planes.

At this point in our discussion we discontinue the use of the glass projection box; you should be familiar with its principle. An **oblique surface** is shown in Figure 3-30. It is a plane surface of any shape that is not perpendicular or parallel to any of the projection planes. The oblique surface does not appear true size in any of the views. To obtain the true size, auxiliary views are required and they are discussed in Chapter 6. Fortunately, oblique surfaces do not occur too frequently in objects, and often it is not too important to have a true view of the oblique surface.

Normal and inclined surfaces can frequently be visualized and sketched without elaborate projection techniques. An oblique surface, however, requires analysis and projection of the endpoints of the surface.

Objects frequently have a number of **curved surfaces.** Material-removal operations and the resulting changes in the shape of the object are shown in Figure 3-31 by a pictorial sketch and a three-view sketch. Let's start with a rectangular solid and machine a cylindrical surface, as shown in Figure 3-31(a). Note the side view where the line stops at the centerline. This type of projection occurs whenever one surface is tangent to another surface.

Now a small hole is drilled part way through the object, as shown in Figure 3-31(b). Note that

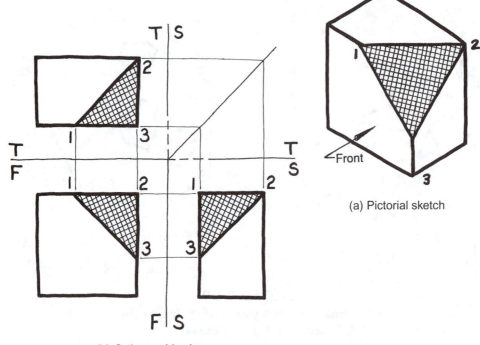

(a) Pictorial sketch

(b) Orthographic views

FIGURE 3-30
Oblique surface.

at the end of the drilled hole a triangle is formed by the end of the drill bit. Hidden or dashed lines are used for the drilled hole in the top and side view because these surfaces are not visible in that view.

The next step is to drill and counterbore two holes for cap screws, as shown in Figure 3-31(c). Note that the bottom of the larger hole is a normal surface and shows a hidden line. Only one hole is shown in the side view because the other one is directly behind it.

Next we can machine another curved surface partway into the object, as shown in Figure 3-31(d). Note how this surface shows up in the top and side view as a rectangle. If this surface had been machined to the end of the part, the long hidden line would have gone to the end of the part and the short hidden line eliminated.

When curves are irregular in shape, such as elliptical, exponential, involute, and so forth, it is necessary to project a number of points and then draw a smooth curve joining the points.

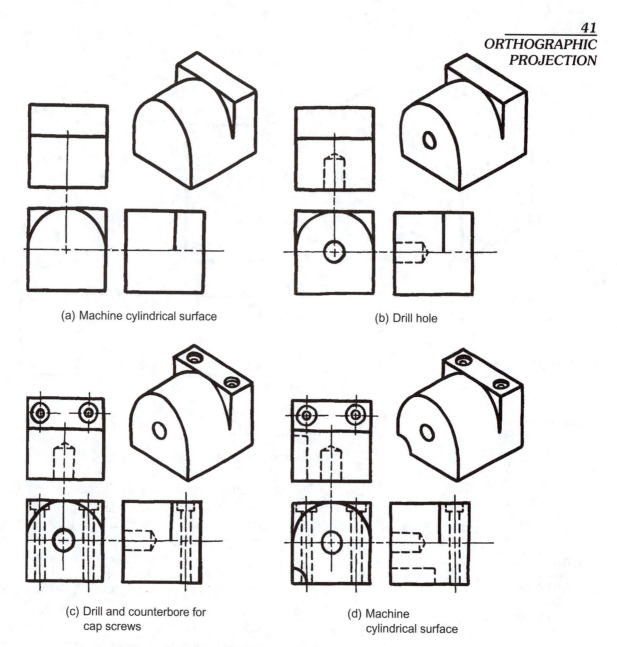

(a) Machine cylindrical surface

(b) Drill hole

(c) Drill and counterbore for
cap screws

(d) Machine
cylindrical surface

FIGURE 3-31
Material-removal operations
and curved shapes.

NUMBERING OF POINTS

Many times in complicated orthographic sketches we need to number points to determine their location in related views. There is no set standard to follow, but once the point is numbered, such as in the pictorial in Figure 3-32, there is only one proper place for it in the orthographic views. In the use of this technique, the point is numbered in two

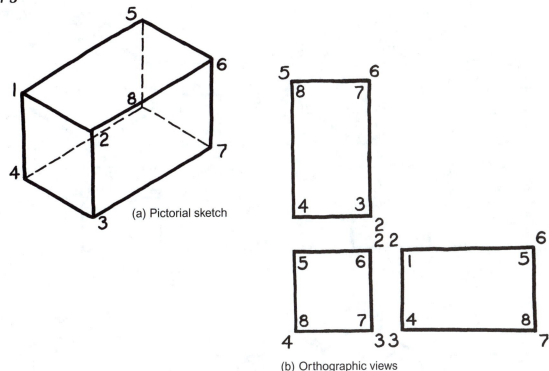

(a) Pictorial sketch

(b) Orthographic views

FIGURE 3-32
Numbering points.

related views, and then located in the third view. Notice that the points will be in line vertically in the front and top views and in line horizontally in the front and side views. The best practice to follow is to use the same number in all views. Letters can also be used to identify points.

The numbers of all visible points are placed outside that view and invisible or hidden ones inside. For example, the point numbered 3 is visible in the front view, so it is located outside the view. In the top view that same point (number 3) is invisible and, therefore, located inside the view.

Let's take a more difficult object and see how the technique works. Suppose that we have the two views shown in Figure 3-33(a). The numbers 1, 2, 3, and 4 have been used to identify the vertical surface on the right. The number 1 is at the top and front of the object; the number 2 is at the top and rear of the object; the number 3 is at the bottom and front of the object; and the number 4 is at the bottom and rear of the object. These four points are projected to the side view as shown by the arrows in Figure 3-33(b) and give the surface 1, 2, 3, 4.

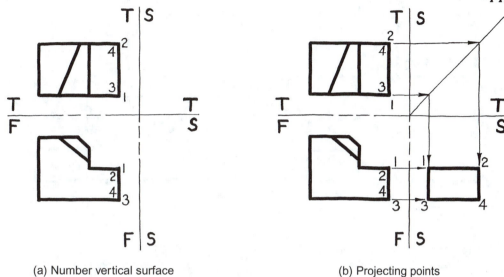

(a) Number vertical surface   (b) Projecting points

FIGURE 3-33
Numbering points.

We can now label another surface, say 5, 6, 7, and 8, the vertical surface on the left. Also we can label 9, 10, which forms a horizontal surface with 1, 2. The projection of these points to the side view is shown in Figure 3-34. Where the lines intersect, the points 5, 6, 7, 8, 9, and 10 are located.

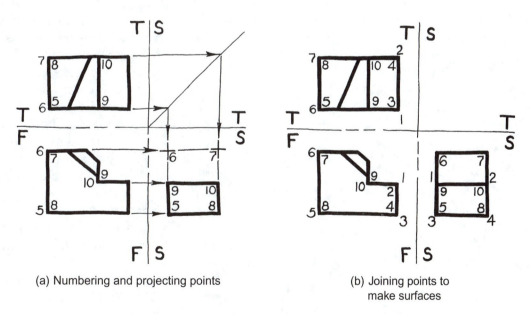

(a) Numbering and projecting points   (b) Joining points to make surfaces

FIGURE 3-34
Numbering points.

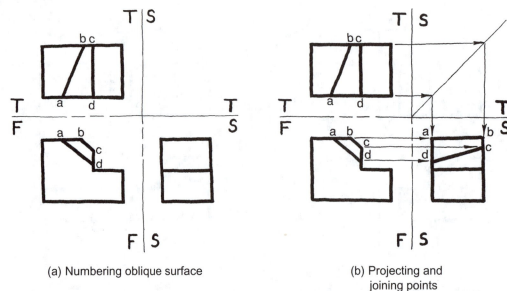

(a) Numbering oblique surface   (b) Projecting and joining points

FIGURE 3-35
Numbering points.

Wherever two points are joined in the front or top view, they must also be joined in the side view. The joined points are shown in Figure 3-34. Note that since points 6 and 10 are not joined in either the front or top views they are not joined in the side view.

Now we have some type of oblique surface that has not been labeled. Let's label the front view with the letters a, b, c, and d, as shown in Figure 3-35(a). For the top view there is only one possible location for points a and b. For c and d there are two possible locations in the top view; however, since d is visible and lower than c in the front view, it must be closest to the projection plane in the top view. The points a, b, c, and d are projected to the side view as shown in Figure 3-35. By joining the respective points together, the complete orthographic projection is obtained.

Figure 3-36 shows a pictorial of the object. Of course, if the pictorial had been given at the beginning, the numbering system would not have been necessary.

Curved surfaces are frequently located by projecting their centerlines. However, it is sometimes necessary to project points on the curved surface and then join the points together. In Figure 3-37(a) we have a cylinder with an inclined surface that will project to the side view as an ellipse. The points 1,

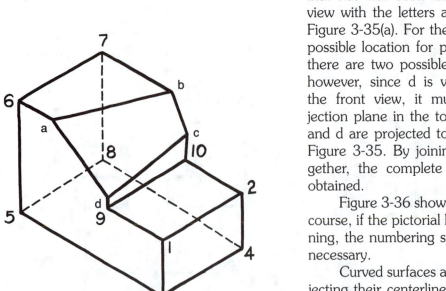

FIGURE 3-36
Pictorial of object

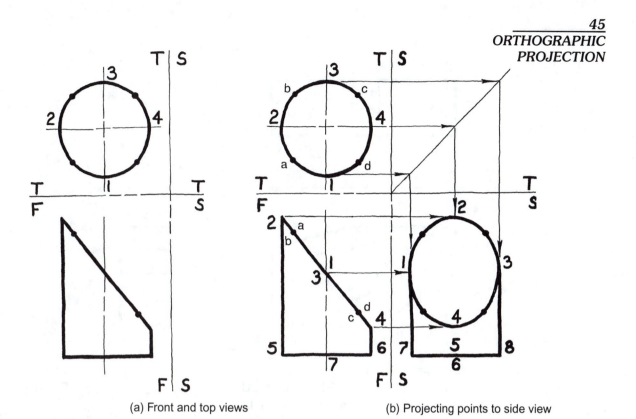

(a) Front and top views   (b) Projecting points to side view

FIGURE 3-37
Projection of a curved surface.

2, 3, and 4 are labeled where the centerlines cross the circle and are projected to the side view by the thin lines with arrows as shown in Figure 3-37(b). Additional points as shown by the letters a, b, c, and d may be necessary to obtain a smooth curve. For clarity the projection of these points is not shown in the figure. The rest of the view can be completed by visualization or projecting the points 5, 6, 7, and 8.

In our discussion we have concentrated on projection to the side view. If the front and side views are given, we could just as easily project the numbers to the top view. Or if the top and side views are given, we could project the numbers to the front view.

PRECEDENCE OF LINES

When two types of lines are coincidental or overlapped one has precedence over the other. A visible line takes precedence over a centerline or hidden line (see a and b in Figure 3-38). A hidden line takes precedence over a centerline (see c).

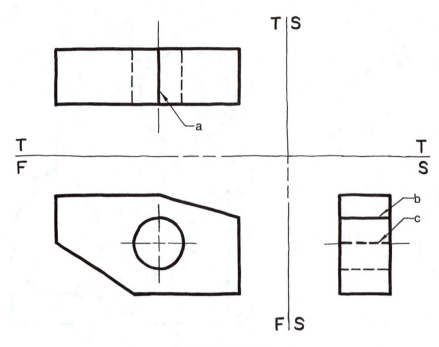

FIGURE 3-38
Precedence of lines.

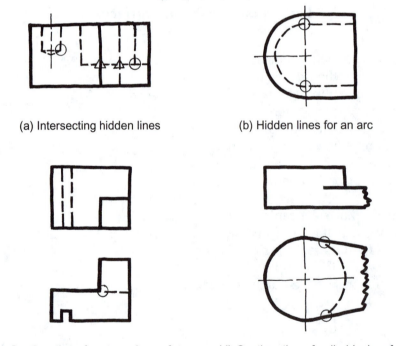

(a) Intersecting hidden lines

(b) Hidden lines for an arc

(c) Continuation of rectangular surface

(d) Continuation of cylindrical surface

FIGURE 3-39
Hidden-line conventions.

HIDDEN-LINE CONVENTION

Certain conventions need to be followed when hidden lines join object lines and other hidden lines. These conventions are illustrated in Figure 3-39. Where hidden surfaces intersect, the hidden lines also intersect, as illustrated by the circles in Figure 3-39(a). Where hidden surfaces do not intersect with other hidden surfaces or visible surfaces, there is no intersection, as illustrated by the triangles in Figure 3-39(a). Hidden lines representing an arc should stop at the endpoints of the arc, with a break for the continuing straight-line part, as illustrated in Figure 3-39(b).

Where a hidden line represents a continuation of a visible surface, a break should be left at the joint with the solid line, as illustrated in Figure 3-39(c). This same principle as it applies to cylindrical objects is illustrated in Figure 3-39(d). When two hidden lines are close together, they are staggered, as illustrated in Figure 3-39(c).

SUMMARY

In summary, one must know or do the following:

1. Orthographic projection means right-angle projection.

2. Orthographic projection uses projection planes.

3. Each set of two adjacent projection planes is perpendicular to each other.

4. The lines of sight (projectors) are parallel to each other and perpendicular to the projection plane.

5. The observer is assumed to be at an infinite distance from the projection plane.

6. A projection is the image of the object on the projection plane.

7. The three most commonly used projections are front view, top view, and side view.

8. The front view is the principal view.

9. Since the observer is at an infinite distance, only one view is observed on the projection plane.

10. Three views are usually necessary to describe an object; however, some objects might only need two or one.

11. Two of the planes of projection appear as an edge in each orthographic view.

12. Folding lines represent the edges or intersections of projection planes.

13. The object is stationary; it is never rotated.

14. A three-view sketch means that the observer has viewed the object from three different directions.

15. Choose the front view to show the most characteristic shape of the object.

16. The object should be in a stable position.

17. View the object to see as many visible lines in the front view as possible.

18. If only two views are needed, show the front and either the top or side.

19. The projection of a point is the foundation of orthographic projection.

20. A normal line is parallel to two projection planes and perpendicular to the other one.

21. An inclined line is parallel to only one projection plane.

22. An oblique line is neither parallel nor perpendicular to a projection plane.

23. A normal surface is parallel to one of the projection planes and perpendicular to the other two.

24. A normal surface is true size in one view and a vertical or horizontal line in the other two views.

25. An inclined surface is perpendicular to one projection plane but inclined to the others.

26. An inclined surface appears as a line in one view and a foreshortened shape in the other two.

27. Oblique surfaces are neither parallel nor perpendicular to any projection planes.

28. The numbering of points can be quite helpful in sketching views.

29. A visible line takes precedence over a hidden line and a centerline, and a hidden line takes precedence over a centerline.

30. Where hidden surfaces intersect, their hidden lines also intersect.

31. Where hidden surfaces do not intersect, their hidden lines do not intersect.

32. A break is left where a hidden line represents a continuation of a visible surface.

PICTORIAL
SKETCHING

CHAPTER

4

Upon completion of this chapter, the student is expected to:

- Know the principles of isometric and oblique sketches.

- Be able to make an isometric sketch from a three-view sketch.

- Be able to make an oblique sketch from a three-view sketch.

- Be able to make an oblique and isometric sketch of a common object.

INTRODUCTION

An orthographic sketch is necessary to adequately represent and describe complicated and intricate parts and machines. This type of sketch is limited, however, to a person who has the special training necessary for its construction and interpretation.

Nearly everyone can read and interpret a photograph of an object. In those cases where intricate detail or interior construction is not needed, the pictorial form of sketch is adequate and very useful. Orthographic sketches show only two dimensions in a view. The various forms of pictorials show three dimensions in a view, thereby giving the object three-dimensional form. For this reason, a pictorial sketch has an important part in graphical representation. It becomes a means of rapid expression and explanation of ideas to the average person.

Our work will be confined to axonometric and oblique types of pictorial sketches, since they are the most convenient to make and the most commonly used.

PRINCIPLES OF AXONOMETRIC PROJECTION

Axonometric projection as illustrated in Figure 4-1 is similar to orthographic projection in that the observer is at an infinite distance, parallel lines of sight are perpendicular to the projection plane, and an image of the object is projected onto the projection plane. One of the differences between the two types of projections is the number of projection planes. Orthographic has three principal projection planes; however, axonometric has only one. Another principal difference is the position of the object. In orthographic projection, the object is in a stable position. In axonometric projection the object is in an unstable position, similar to a cube suspended on one point on a table.

There are three different types of axonometric projection. The type is determined by the position of the object. Since the object is rotated and tilted, a different relationship of the angle between the axes and the length along the axes is obtained.

Isometric projection occurs when the object is rotated and inclined to the projection plane in such

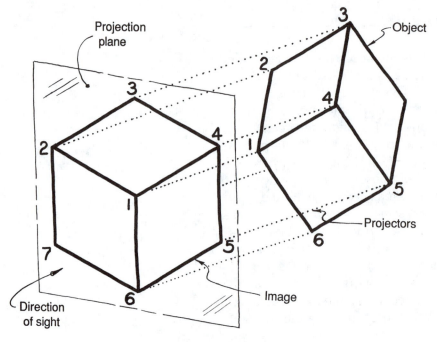

FIGURE 4-1
Axonometric projection.

a manner as to produce equal angles among the three axes and equal lengths along each axis. Figure 4-2(a) illustrates an isometric projection of a cube. Dimetric projection (Figure 4-2b) occurs when the object is rotated and inclined to the projection plane in such a manner as to produce only two axes angles and two axes lengths that are equal. Trimetric projection (Figure 4-2c) occurs when the object is rotated and inclined to the projection plane in such a manner that none of the

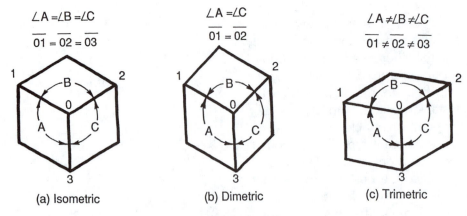

FIGURE 4-2
Types of axonometric projection.

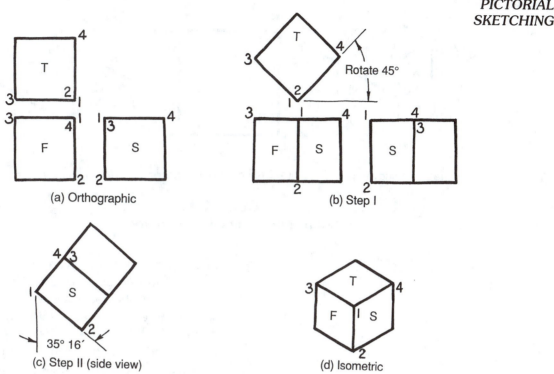

FIGURE 4-3
Isometric projection by revolution.

three axes angles or lengths is equal. Because of the limited use of dimetric and trimetric in sketching, we will only study the isometric projection.

Isometric projection is the most widely used, since the same scale may be used on all axes. Figure 4-3 shows the development of an isometric projection. A three-view orthographic sketch of a cube is illustrated in Fig. 4-3(a). Assume that the cube is placed flat on, say, a level tabletop, with one surface parallel to the front projection plane that is perpendicular to the tabletop. To produce the isometric projection, rotate the cube to an angle of 45° with the front plane, viewed from the top as in Fig. 4-3(b). The cube remains flat on the table. Next, tip the cube forward until the front corner 1, 2 makes an angle of 35° 16′ with the front plane as viewed from the side as in Fig. 4-3(c). This position gives the isometric projection of the cube as shown Fig. 4-3(d) and establishes the isometric axes. One plane of projection, the front projection plane is used. The object is rotated and tipped to show the three surfaces of the cube in one view.

The resulting isometric axes are equal and at an angle of 120° to each other, as shown in Figure 4-4(a). For simplicity, one of the axes is established

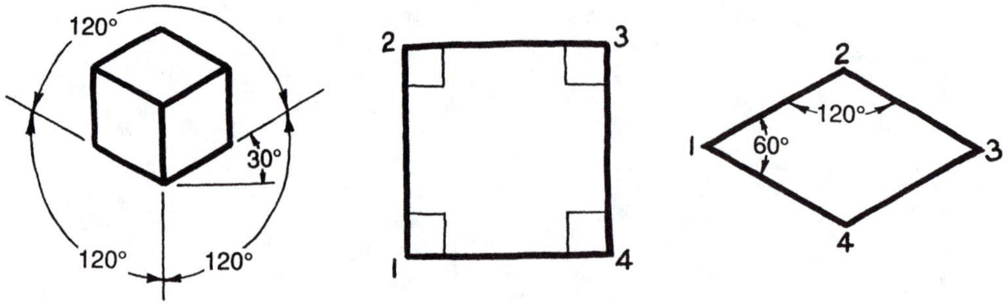

FIGURE 4-4
Square in isometric projection.

(a) Isometric axes

(b) Orthographic square

(c) Isometric square

in a vertical position, which requires the other two axes to form an angle of 30° with the horizontal. Figure 4-4(b) shows a one-view sketch of a square surface. At each corner is a 90° angle, as shown at corners 1, 2, 3, and 4. In Figure 4-4(c) the square is shown in isometric form and the angles at corners 1 and 3 are at 60° and the angles at corners 2 and 4 are at 120°. In the orthographic sketch, the diagonals of the square are equal. In the isometric, the diagonals are unequal, neither being the true length of the diagonal. *Thus measurements can be made only along or parallel to an isometric axis.*

Since the object has been rotated and inclined at an angle with the projection plane, it will appear foreshortened when projected on that plane. A true isometric projection is foreshortened by 81.6%, as shown in Figure 4-5(a). Special scales

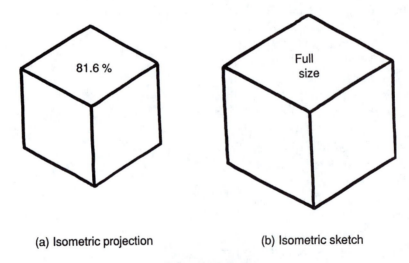

(a) Isometric projection

(b) Isometric sketch

FIGURE 4-5
Isometric and ordinary scales.

are available to construct an isometric projection. However, in practical work, an isometric sketch is constructed full size as shown in Figure 4-5(b). Since the three isometric axes are at 120° to each other, the difference between an isometric projection and an isometric sketch is one of size, the isometric sketch being larger than a true projection.

ISOMETRIC SKETCHING TECHNIQUES

The most important sketching technique is the use of isometric paper. This paper must be turned so that the sloping lines make an angle of 30° with the horizontal axis, as shown by Figure 4-6. Lengths may be estimated by the grid on the paper. The paper is very helpful for sketching rectangular, sloping, or circular surfaces. Sketch lines should be dark and prominent. As noted in Chapter 1, special sketching pads of transparent paper are available that allow an isometric grid to be placed underneath the transparent paper. The isometric grid is visible through the transparent paper, and the resultant sketch has a more pleasing appearance without the isometric grid showing. Note the method of sketching the fillet; a round would be sketched similarly.

Figure 4-7 illustrates the isometric box construction technique. When working with objects with inclined surfaces, or cutouts, it is best to use the box construction method. To do this, merely

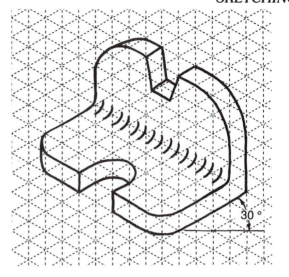

FIGURE 4-6
Isometric sketch using isometric paper.

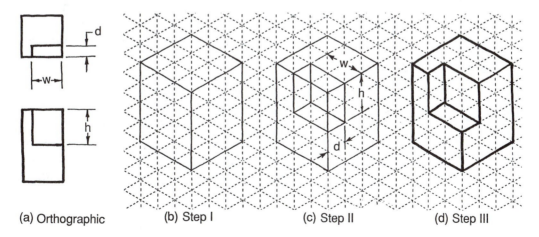

(a) Orthographic   (b) Step I   (c) Step II   (d) Step III

FIGURE 4-7
Box construction method.

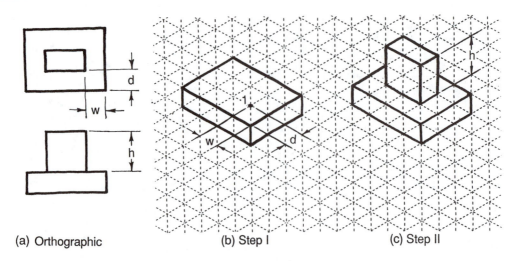

(a) Orthographic   (b) Step I   (c) Step II

FIGURE 4-8
Block method.

show the outline of a box totally enclosing the object (step I); then make measurements along the edges of the box to locate the cutout corners or surface lines. Dimensions *w, h,* and *d* illustrate this technique (step II). The box should be developed using construction lines (3H-grade lead) and then sketched with prominent lines (HB-grade lead or a fiber pen) after the configuration has been established (step III).

Another technique is to build the piece up one block at a time, as shown in Figure 4-8. In isometric sketching, vertical measurements are made only on or parallel to the vertical axis. Horizontal measurements are made only on or parallel to the horizontal axes. In using the block method, first sketch the lower block. To locate corner 1 (step I), we must make two horizontal measurements, *w* and *d.* We can now sketch height *h* and complete the upper block as shown in step II.

To sketch inclined surfaces, it is necessary to locate the end points of the surface, as illustrated in Figure 4-9 using the box-construction method. Step I locates the points 1 and 4 using one vertical measurement *a* and one horizontal measurement *b,* step II locates the points 2 and 3 using one horizontal measurement *c,* and step III completes the isometric by joining the points.

Projection of features onto sloping surfaces requires the projection of sufficient points to outline the surface. In Figure 4-10 we have a square hole opening out onto an inclined surface. To

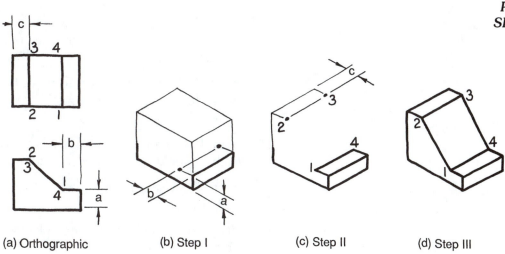

(a) Orthographic   (b) Step I   (c) Step II   (d) Step III

FIGURE 4-9
Inclined surfaces.

show the hole opening on the inclined surface, construct a box enclosing the object in both the orthographic and isometric.

Step I. Sketch in the projection of the opening on the end of the box 1234 in isometric.

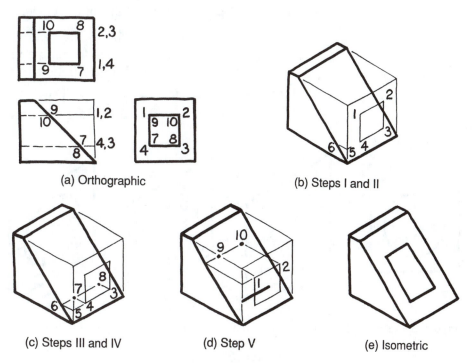

(a) Orthographic   (b) Steps I and II

(c) Steps III and IV   (d) Step V   (e) Isometric

FIGURE 4-10
Projection onto inclined surfaces.

Step II. Project points 3 and 4 to the corner of the box, point 5. Then project to point 6 on the sloping surface.

Step III. Project point 5 to point 6 on the sloping surface. Sketch a horizontal line on the sloping surface from 6.

Step IV. Project points 3 and 4 horizontally back to the sloping surface to locate points 7 and 8.

Step V. Use the same method to locate points 9 and 10.

Step VI. Sketch in the complete outline of the object.

When using isometric paper, the grid lines make the construction much simpler.

Figure 4-11 illustrates the proper procedure to follow in constructing circles in isometric. It is known as the four-center ellipse method. Given is a square in orthographic with a circle on the surface of the square.

Step I. Lay out the square in isometric.

Step II. Lay out the two centerlines of the circle on this surface.

Step III. On the centerline, lay out the four points for the radius of the circle.

Step IV. Parallel to the centerline, sketch lines through these points, forming an isometric square that would totally enclose the circle.

Step V. From corners 1 and 2 of the parallelogram, sketch lines to the midpoints of the opposite

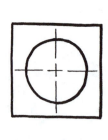 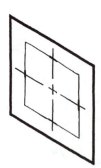 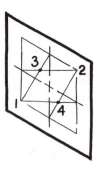 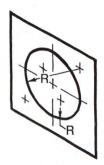

(a) Orthographic   (b) Steps I, II, III, and IV   (c) Step V   (d) Step VI

FIGURE 4-11
Isometric circle construction.

sides. These lines intersect at points 3 and 4, which are the center of the two end radii. Points 1 and 2 are the two centers of the other radii.

Step VI. Complete the circle by sketching all four arcs.

When using isometric paper, many of these construction lines are available, provided the circle center is located properly. With practice one should be able to adequately sketch a circle in isometric with just the use of the isometric grid. Many designers use an ellipse template.

The preceding description is applied to the front surface of a cube. For the top and side surfaces the construction is the same, but the appearance is different, as shown in Figure 4-12. It is important to notice that the long diagonal of the parallelogram (the major axis of the ellipse) will slope upward to the right in the side surface, upward to the left in the front surface, and be horizontal in the top surface.

Angles in isometric are treated similarly to sloping surfaces. Figure 4-13 shows angles laid off in orthographic and the proper procedure for laying off angles in an isometric sketch. Distances along the isometric axes must be obtained in the orthographic view and measured along the isometric axes in the isometric sketch. The line, representing the angle, may then be drawn to this point. Angles may not be measured by their degrees in isometric sketches as these angles are bounded by nonisometric lines and are not true measurements.

FIGURE 4-12
Isometric circles on different surfaces.

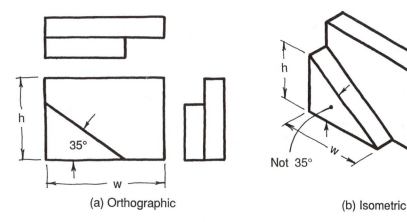

(a) Orthographic

(b) Isometric

FIGURE 4-13
Angles in isometric.

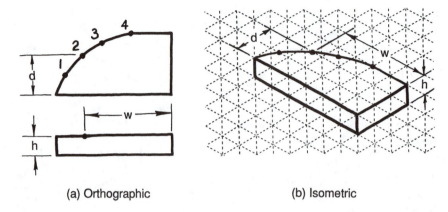

| (a) Orthographic | (b) Isometric |

FIGURE 4-14
Irregular shapes in isometric.

Irregular shapes are treated as if they were a series of points with each point requiring three measurements in the isometric sketch. The technique is illustrated in Figure 4-14. The distances are obtained from the orthographic view, and measurements are made along or parallel to the isometric axes to establish a series of points. By joining these points, a smooth curve is constructed. Enough points should be located to ensure sketching a curve representing the contour shape of the object. The location of point 2 is illustrated by the distances *w*, *h*, and *d*. The plotting of points would also have been necessary for a round hole or an inclined surface rather than the square hole of Figure 4-10.

Because of the shape and location of some objects, a reversed isometric axis is sometimes used. Figure 4-15(a) shows the reverse position of the isometric axes in which we are looking up from the bottom of the object rather than down from the top. In Figure 4-15(b) the basic cube is illustrated, and Figure 4-15(c) is a sketch of an object.

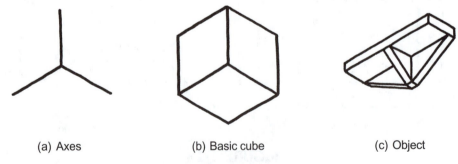

| (a) Axes | (b) Basic cube | (c) Object |

FIGURE 4-15
Reversed isometric axes.

The orientation of the object should be accomplished to maximize the number of visible lines. Hidden lines are only used if the object is not adequately described by the visible lines. If necessary, the axes can be oriented so that none are vertical; however, they must be 120° to each other.

OBLIQUE SKETCHING

Oblique sketching is a form of pictorial representation that offers the advantage of easily illustrating objects having irregular shapes on one surface only. Oblique projection is not a true projection. While the observer is considered to be at an infinite distance and the lines of sight are parallel, they are at an angle to the projection plane rather than perpendicular. All surfaces that are parallel to the projection plane remain in that configuration while the rest of the object is rotated and tilted.

Since oblique is not a true projection, the object is actually distorted. Figure 4-16(a) shows two orthographic views of an object. In Figure 4-16(b) (front and top views) the object is rotated; however, surfaces F, M, and R, which are parallel to the projection plane, remain parallel. This is analogous to taking a rectangle and converting it to a parallelogram. The object is tilted upward, keeping surfaces F, M, and R parallel to the projection plane as shown in Figure 4-16(c), which is a side view. The completed oblique sketch is shown in Figure 4-16(d).

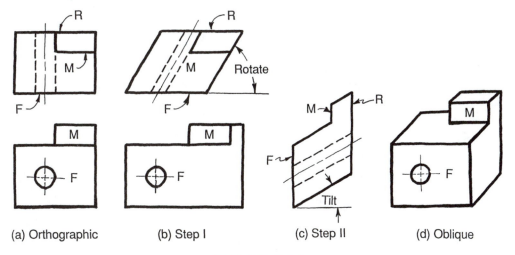

(a) Orthographic   (b) Step I   (c) Step II   (d) Oblique

FIGURE 4-16
Oblique projection.

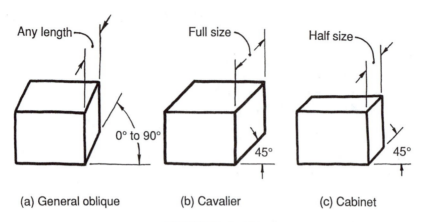

FIGURE 4-17
Types of obliques.

For the general oblique sketch, the angle of the receding axis can be varied from 0° to 90° and the length to any scale we choose, as shown in Figure 4-17(a). Since oblique sketching is not a true projection, these liberties can be taken. Two specific classifications of oblique sketches are constructed with a receding axis of 45° and are shown in Figure 4-17(b) and (c). These are classified as **cavalier** and **cabinet** sketches. The scale on the receding axis of the cabinet is half size and the cavalier scale is full size. By varying the angle of the receding axis, we can accent different surfaces. By changing the scale along the receding axis, we can lengthen or shorten the appearance of the object.

Figure 4-18 shows a comparison of a cylinder sketched in both isometric and oblique. Notice the distortion that we get in both projections. The isometric piece looks flattened rather than cylindrical,

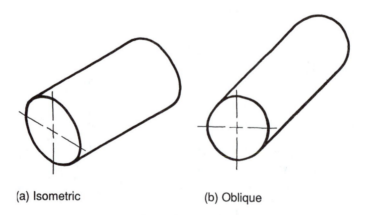

FIGURE 4-18
Comparison of isometric
and oblique sketches.

and the oblique seems to get larger toward the far end of the piece. Oblique sketching has its purpose, but it is not the answer to all types of pictorial sketching, any more than isometric is.

Since the diagonal of a square grid is at 45°, it is best to use a 45° angle for the receding axis. The cabinet style usually has a more pleasing appearance than the cavalier, as shown in Figure 4-19. Therefore, cabinet-style oblique sketches are recommended and discussed in the sections that follow.

The first and most important step in the construction of a cabinet-style oblique sketch is the selection of the front face. Circular or irregular surfaces of the object should be selected for the front face. Since the front surface of the object is parallel to the projection plane, it is a true orthographic front view and is constructed in the same manner. This fact is the basic reason for using an oblique sketch rather than an isometric sketch. Figure 4-20 illustrates the importance of the selection of the front face. The sketch shown in 4-20(a) is the correct one, since it is the easiest to sketch and has the most pleasing appearance. It is more difficult to construct the circles when the object is positioned as shown in Figure 4-20(b) and (c).

Figure 4-21 shows the method of transferring centers along the receding axis. The circle is sketched in the front view; then, along the 45° axis of the center, a new center is established at the rear of the object equal to one-half the length of the object (step 1). Since the actual length is 4 squares, the length of the receding axis is 2 squares. Mea-

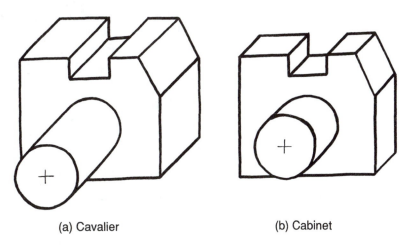

(a) Cavalier   (b) Cabinet

FIGURE 4-19
Comparison of cavalier and
cabinet projections.

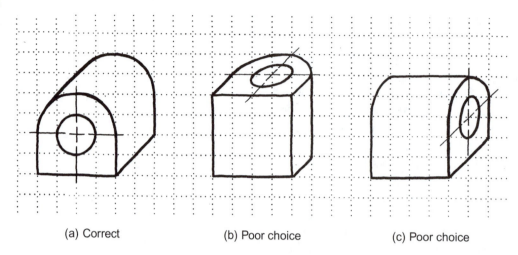

(a) Correct (b) Poor choice (c) Poor choice

FIGURE 4-20
Selection of the front face.

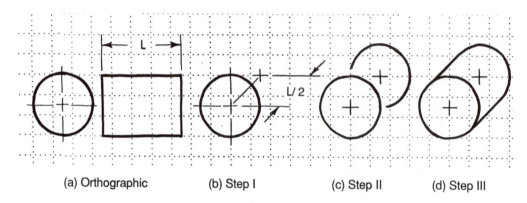

(a) Orthographic (b) Step I (c) Step II (d) Step III

FIGURE 4-21
Transfer of centers along an axis.

surement can be made by ruler or a piece of paper with tic marks at 2 squares. The necessary portion of the rear circle is sketched in step II, and the sides of the cylinder are sketched tangent to the circles as shown in step III.

Steps for a more complicated object are shown in Figures 4-22 and 4-23. Figure 4-22 shows the orthographic views and steps I and II. The first step is to sketch the large circle and the small circle. To locate the center of the small circle, dimension a is laid off on the 45° receding axis and then dimension b on the horizontal axis. Note that in the oblique sketch dimension a is one-half of dimension a in the orthographic view. Step II requires the location of the rear center of both cylinders and the sketching of these circles. Both dimensions c and d are laid off along the receding axis a distance equal to one-half of dimensions c and d in the orthographic view.

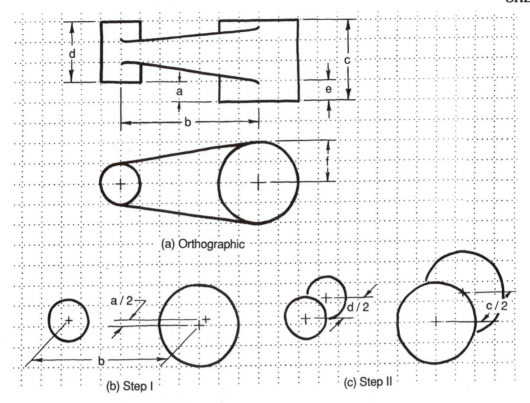

FIGURE 4-22
Steps for an oblique sketch.

The location of the web of the casting is shown in Figure 4-23. Step III is the location of the eight tangent points of the web with the two cylinders. The location of one tangent point is shown by the dimensions *e* and *f*. After the other seven tangent points are located, the tangent lines are sketched. Actually, only five additional tangent points are necessary since one of the lines is hid-

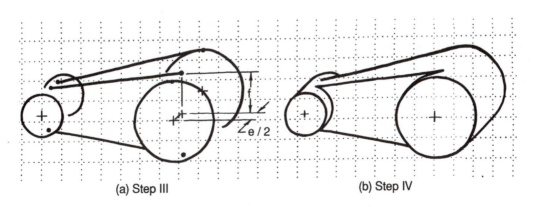

FIGURE 4-23
Steps for an oblique sketch.

den. Step IV requires the two circles at the web to be constructed and the entire sketch completed.

SUMMARY

In summary, one must know or do the following:

1. The three types of pictorial projection are axonometric, oblique, and perspective.

2. Isometric, diametric, and trimetric are the three types of axonometric.

3. An isometric projection is 81.6% of its full size.

4. In isometric sketching the object is rotated so that the angles between the three axes are equally spaced at 120°.

5. Measurements on all three axes use the same scale in isometric.

6. For isometric sketching, the object is turned to see as much detail as possible.

7. The distances (height, width, and depth) are transferred from the orthographic sketch along or parallel to the isometric axes.

8. Oblique is not a true projection because the line of sight is not perpendicular to the projection plane.

9. Oblique is beneficial in showing objects with irregular or circular shapes on one surface and on surfaces parallel to that surface.

10. Cavalier and cabinet oblique have 45° receding axes.

11. Cabinet oblique differs from cavalier in that the receding axis scale is one-half rather than full size.

12. Cabinet oblique has a more pleasing appearance than cavalier.

SECTIONAL VIEWS

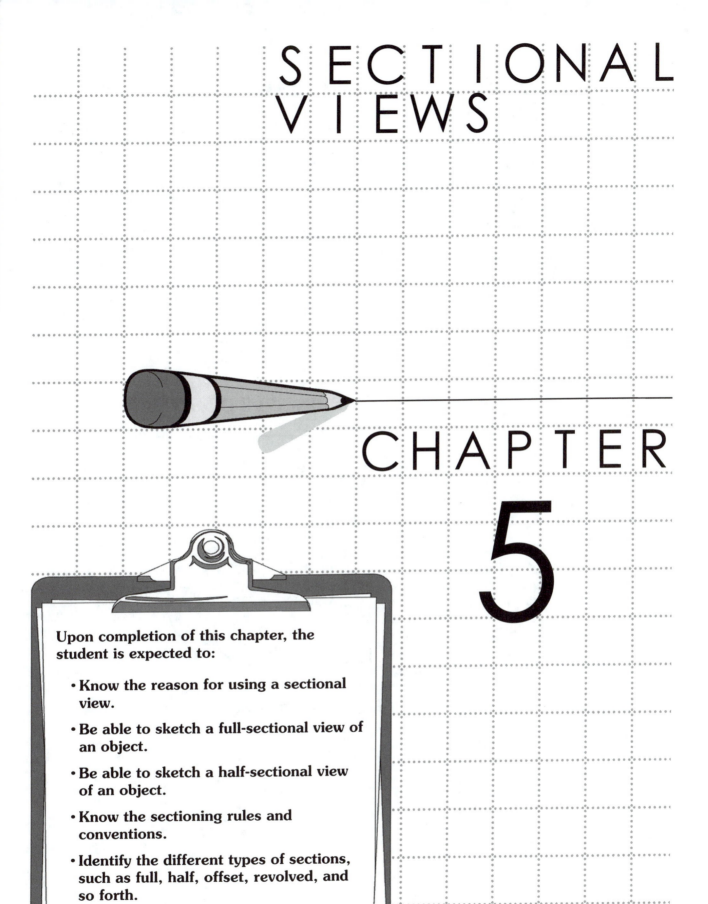

CHAPTER

5

Upon completion of this chapter, the student is expected to:

- Know the reason for using a sectional view.

- Be able to sketch a full-sectional view of an object.

- Be able to sketch a half-sectional view of an object.

- Know the sectioning rules and conventions.

- Identify the different types of sections, such as full, half, offset, revolved, and so forth.

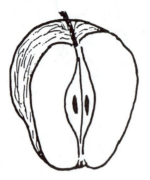

FIGURE 5-1
Sliced apple.

INTRODUCTION

Sectional views are used to describe the interior construction and detail of parts and machines. The sectional view is particularly helpful in describing large or intricate pieces, such as an engine block, electric motor housing, and other objects where the exterior orthographic view becomes difficult to interpret because of so many hidden lines.

The concept of sectional views can be illustrated by slicing in half an apple. Figure 5-1 represents an apple that has been cut in half and one part removed. The interior features, including the seeds, core, and stem, can be clearly seen. This is not possible in a regular orthographic view without numerous hidden lines.

PRINCIPLE OF SECTIONAL VIEWS

The principle of taking sectional views is shown in Figure 5-2. The orthographic view of the object is shown at (a) and the cabinet oblique view of the ob-

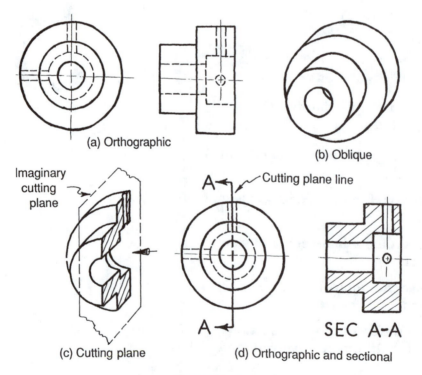

(a) Orthographic

(b) Oblique

(c) Cutting plane

(d) Orthographic and sectional

FIGURE 5-2
Sectional view theory.

ject is shown at (b). An imaginary cutting plane is passed through the center of the object and one half of the object removed, as illustrated at (c). An arrow indicates the direction of sight, which is perpendicular to the cutting plane. The rear half of the object is behind the cutting plane. Where the cutting plane comes in contact with the object, the surface is cross-hatched. The orthographic front view and the sectional view as they would appear on the sketch are shown at (d). A **cutting plane** line is placed on the orthographic view to show the location of the cutting plane. At the end of the cutting plane line, arrows are drawn to indicate the direction of sight. Since more than one sectional view may be required to describe an object, letters are used for identification. In this case the letter A is used. The thickness of the cutting plane line is the same as a visible line. Construction of the cutting plane line is by a series of two short dashes and one long dash. The short dashes are approximately 3 mm (0.12 inch) long with a space between of approximately 1 mm (0.04 inch), and the long dashes can vary from 20 to 40 mm (0.8 to 1.6 inches). The cutting plane line is placed on the orthographic view that represents the cutting plane as an edge.

The sectional view is shown next to the orthographic view. Note the crosshatched areas which show the exposed areas, and the label SEC A-A at the bottom of the view, which identifies that particular sectional view.

A sectional view is a special type of orthographic view that is better able to show the interior features of a complicated object. The imaginary cutting plane is identical to the imaginary reference plane of an orthographic view except that it is placed against the object after an imagined portion of that object is removed. Therefore, the only difference between the two types of views is in their technique of representation. Sectional views are accompanied by all necessary orthographic views.

The section line technique of the crosshatched area is shown in Figure 5-3. Section lines are thin and are located approximately 3 mm (0.12 inch) to 5 mm (0.20 inch) apart. This spacing is not measured, but is estimated. Larger sketches permit greater spacing; smaller sketches permit closer spacing. The spacing should not be so close or so heavy as to make the crosshatched area look dark in comparison to the rest of the sketch. The outline

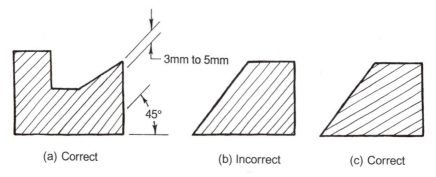

(a) Correct   (b) Incorrect   (c) Correct

FIGURE 5-3
Crosshatched.

or visible lines should stand out as the most prominent lines.

As a general rule, section lines are sketched at an angle of 45°; however, when the lines parallel the side of the object, the appearance is not too pleasing, as shown in Figure 5-3b. In such cases an angle of 30° as shown at (c) or 60° is more appropriate. The particular style of section lines illustrated in Figures 5-2 and 5-3 is the one for general use. It is also the style for cast iron. Section lining symbols for other materials are shown in Figure 5-4. It is usually unnecessary to sketch these different material symbols since the exact material must be specified in

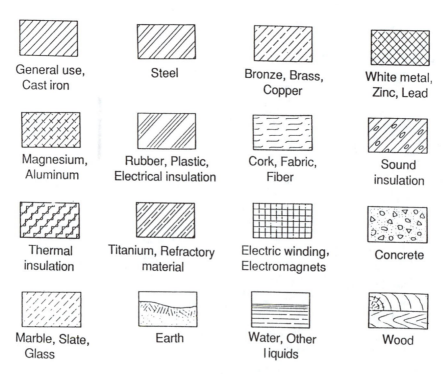

General use, Cast iron	Steel	Bronze, Brass, Copper	White metal, Zinc, Lead
Magnesium, Aluminum	Rubber, Plastic, Electrical insulation	Cork, Fabric, Fiber	Sound insulation
Thermal insulation	Titanium, Refractory material	Electric winding, Electromagnets	Concrete
Marble, Slate, Glass	Earth	Water, Other liquids	Wood

FIGURE 5-4
Symbols for section lining.

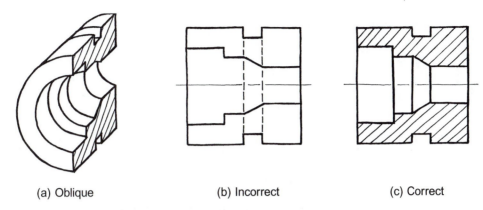

(a) Oblique   (b) Incorrect   (c) Correct

FIGURE 5-5
Visible and invisible lines behind
the cutting plane.

precise terms. For this reason and to save time, the cast iron or general symbol is used on detail sketches. It may, however, be necessary to use the symbols for assembly sketches to show the different parts more effectively.

An important principle in the sketching of sectional views is to show all visible edges behind the cutting plane. The beginner will sometimes omit the visible lines behind the cutting plane as illustrated by the three missing visible lines in Figure 5-5(b). We also have the poor practice of showing hidden lines in the sectional view. Only in special cases where hidden lines would save another view or lend clearness to the sketch are they shown.

Figure 5-5(c) shows the proper technique to follow in the construction of a sectional view. Notice that no hidden lines are shown and that all visible lines behind the cutting plane are shown.

CONVENTIONAL REPRESENTATION

Conventional representation or conventional practice is the term used to describe situations where the true projection is not sketched because it is difficult to accomplish and is not as easy to read. Figure 5-6 shows two sectional views of an object. The sectional view at (b) is a true projection, but is poor practice. The sectional view at (c) is the conventional presentation, and it is much easier to read and quicker to sketch. Conventional representation should be followed in making sectional views as well as the orthographic views described in Chapter 3.

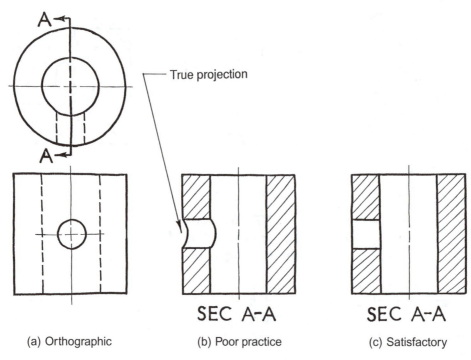

FIGURE 5-6
Conventional representation.

The handling of solid parts and ribs or webs in section can result in a confused representation if a true projection is used. The general rule is that ribs, webs, spokes, and so forth are not cross-hatched in sectional views. The omission of cross-hatching the rib or web is to avoid the impression of a solid piece. A sectional view of a solid piece is shown in Figure 5-7(a). In Figure 5-7(b) we have the ribs in section. A true projection of the ribs in the sectional view would result in the same view as the solid piece. Therefore, it is necessary to depart from a true projection and use the conventional representation shown.

Figure 5-8 shows the effect of webs in section. To properly convey to the reader the interior features, a conventional representation as shown at (b) is used rather than the true projection as shown at (c). Note the imagined position of the cutting plane line in the orthographic view as shown at (a) to aid in visualizing the correct method of constructing the sectional view. It is also impor-tant to note that the small holes are considered to be rotated.

The illustration in Figure 5-9(a) shows the cor-rect handling of a solid-disk-type center of an ob-ject. From this sectional view, one would expect to find a solid center to the object, which it has. The

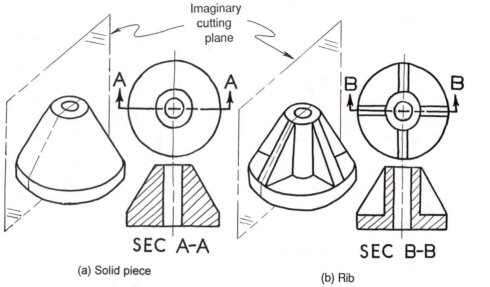

SEC A-A

(a) Solid piece

SEC B-B

(b) Rib

**FIGURE 5-7
Ribs in section.**

illustration at (b) represents a spoke wheel, and the spokes in the section view are not crosshatched. If they had been crosshatched, the impression of a solid piece would be conveyed.

As a general rule, conventional practices are used to adequately convey the interior shape of the object. It is also conventional practice to revolve external features and show their true shape in the sectional view.

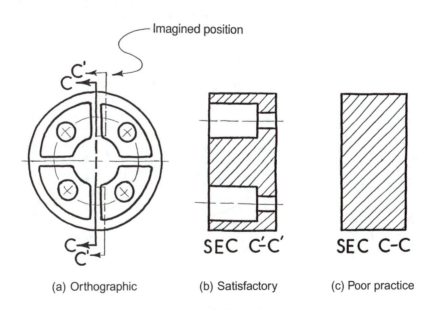

(a) Orthographic

(b) Satisfactory

(c) Poor practice

**FIGURE 5-8
Webs in section.**

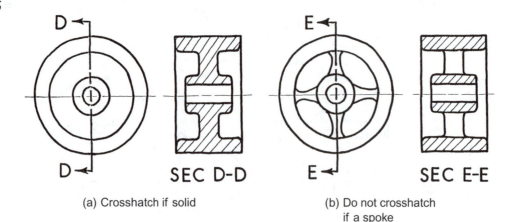

D → D

SEC D-D

(a) Crosshatch if solid

E → E

SEC E-E

(b) Do not crosshatch
if a spoke

FIGURE 5-9
Flanges and spokes in section.

FULL AND HALF SECTIONS

Our discussion of sectioning has been concerned with the full-sectional view, which is the most common type. Another common type is the half-sectional view. Before discussing the half-sectional view, let's review the full section.

Figure 5-10 reviews the principle of taking a sectional view. An imaginary cutting plane is passed through the center of the object, and the part of the object in front of the plane is assumed to be removed as shown at (a). The cutting plane line, which is the edge view of the cutting plane, is located in the front orthographic view. Arrows at the ends of the line indicate the direction of sight as

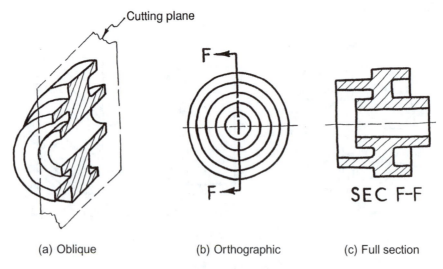

Cutting plane

F → F

SEC F-F

(a) Oblique

(b) Orthographic

(c) Full section

FIGURE 5-10
Full section.

shown at (b). Where the cutting plane comes in contact with the object, section lines are sketched in the sectional view. All visible lines behind the cutting plane are shown; however, hidden lines are not shown unless needed for clarity. The sectional view is identified by the letter F, as is the cutting plane line.

Half-sectional views are used with symmetrical objects to show both the interior and exterior views as shown in Figure 5-11a. The cutting plane passes halfway through the object, and one fourth of the object is assumed removed. Arrows indicate the direction of sight. They point in the direction the observer is looking, and therefore both must point in the same direction. This makes one of the arrows have a different appearance than the other one as shown at (b). Again, the sectional view must be identified and the cutting plane line shown. Hidden lines are normally not shown unless needed for dimensioning or clarity.

The object in Figure 5-11 is the same object as given in Figure 5-10; therefore, the differences between the two types of sectional views can be compared. Both the interior and exterior shapes of the object are clearly shown by the half-sectional view as illustrated in Figure 5-11c. Note that two of the vertical lines of the exterior portion and one of the vertical lines of the interior portion stop at the center line. This situation is quite common and makes visualization of half-section views difficult.

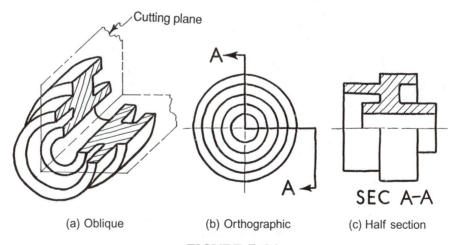

(a) Oblique   (b) Orthographic   (c) Half section

FIGURE 5-11
Half-section.

OTHER TYPES OF SECTIONAL VIEWS

There are other types of sectional views that you should be familiar with and be able to identify. These are offset, revolved, removed or detailed, broken and partial, and assembly. Each of these is applicable in particular cases and is discussed in the information that follows.

Offset

To convey the interior features of an object adequately, it may be necessary to change the direction of the cutting plane line. When this is done, we have an **offset section,** as illustrated in Figure 5-12. The section is identified by arrows and letters, and the line is offset according to the desired detail. The sectional view does not contain the change of direction of the cutting plane line. It is important in an offset section to make the cutting plane line dark and visible.

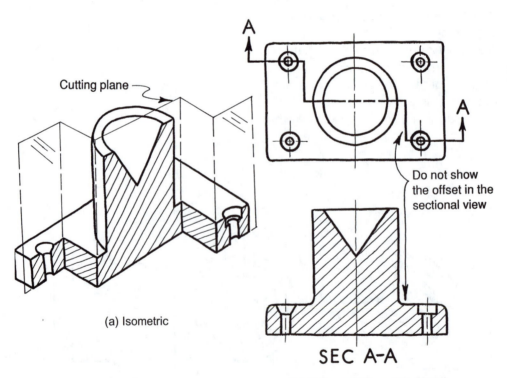

Cutting plane

Do not show
the offset in the
sectional view

(a) Isometric

SEC A-A

(b) Orthographic and sectional

FIGURE 5-12
Offset section.

Revolved

To save the construction of a full-sectional view or another orthographic view, a **revolved section** is used as shown in Figure 5-13. The sectional view is produced by a cutting plane that is revolved 90° about an axis at the center of the revolved section as shown at (a). Lines from the exterior view of the object should not cross the sectional view, as shown at (b). In some cases, such as the extrusion at (c), break lines are used. Two advantages of revolved sections are the elimination of the cutting plane line and of the sectional view designation. In addition, the sketch is frequently easier to understand and a view is saved.

Removed or Detailed

Removed or detailed sections are used to show the shape and construction of the object at a specified location. This type of sectional view is similar to revolved sections in that its purpose is to show the cross-sectional area. These removed sections are often sketched to enlarged scale to permit clear interpretation of the part construction. They must clearly show where the section was taken and the direction of sight. A number of removed sections in sequence can be used to show the change in shape of an object. An example of a removed or detailed section is shown in Figure 5-14. The sectional view can be located anywhere on the sketch in contrast to locating it as one of the principal views as shown in the figure. An excellent example of this type of sectional view would be a number of removed or detailed sectional views to describe the shape of an airplane wing or fuselage. Note the use of section lines with break lines on the right side of the sketch to illustrate a solid rod. This technique could also be used for a tube be showing the inside diameter.

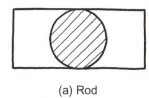

(a) Rod

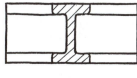

(b) I-beam

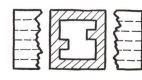

(c) Extrusion
with break lines

FIGURE 5-13
Revolved section.

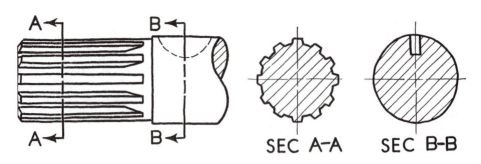

FIGURE 5-14
Removed or detailed section.

Broken and Partial

If only a small portion of an object is to be shown, a **broken and partial section** may be desirable. By this means, important interior construction of a small portion may be shown without full sectioning of the part or another view. An example of usage would be cross-drilled holes for bearings, locking setscrew holes, and threaded sections. Figure 5-15 illustrates an excellent example of use of this type of sectional view.

Assembly

Our last type of sectional view is the **assembly section,** which is shown in Figure 5-16. An assembly section is used to show the working relationship of parts in the unit and serves as a place to call out identification numbers and names. All areas of the same detail part should have the same section lines sloping at the same angle. Where the cutting plane passes along the center of bolts, nuts, pins, keys, spokes, ribs, or webs, these items should not be crosshatched in the sectional view.

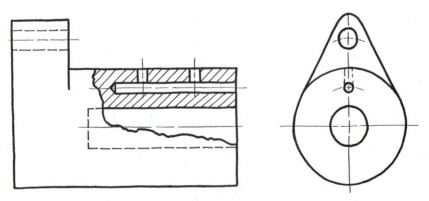

FIGURE 5-15
Broken and partial sections.

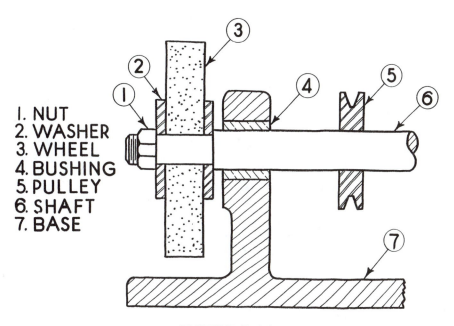

1. NUT
2. WASHER
3. WHEEL
4. BUSHING
5. PULLEY
6. SHAFT
7. BASE

FIGURE 5-16
Assembly section.

They should, however, be crosshatched if the cutting plane passes perpendicular to their axis. Section lines of adjacent parts should be crosshatched in opposite directions. The identification number permits reference to the detail drawing or catalog illustration for full information on the part. Note the use of different material symbols for the crosshatching.

SUMMARY

In summary, one must know or do the following:

1. Sectional views are produced by an imaginary cutting plane passing through the object.

2. The sectional view is the portion of the object behind the cutting plane.

3. The observer is at an infinite distance.

4. A sectional view is used to show the interior shape of an object.

5. Only material in contact with the cutting plane is crosshatched.

6. The arrows on a cutting line point in the direction in which the observer is looking.

7. The cutting plane line and sectional view must be identified.

8. Crosshatch lines should be spaced approximately 3 to 5 mm (0.12 to 0.20 inch) apart.

9. The crosshatched area should not appear dark on the sketch.

10. Hidden lines are not shown in the sectional view.

11. All visible lines behind the cutting plane are shown.

12. A sectional view is accompanied by all necessary orthographic views.

13. A half-section is used only with a symmetrical object.

14. A half-section shows both the interior and exterior of the object.

15. Conventional representation, which is not a true projection, is used to make the sketch easier to read and much quicker to construct.

16. In an offset sectional view the cutting plane line changes direction.

17. A revolved sectional view is quicker to construct and easier to read than a full sectional view.

18. Removed or detailed sectional views are used to show the change in shape of an object such as an airplane wing.

19. Interior construction of a small portion is shown by a broken and partial sectional view.

20. An assembly sectional view shows the working relationship of the parts.

AUXILIARY VIEWS

CHAPTER

6

Upon completion of this chapter, the student is expected to:

- Know the reasons for using auxiliary views.

- Understand the concept of primary auxiliary views.

- Be able to obtain the true length of an oblique line.

- Be able to obtain the true size of an inclined surface.

- Understand the concept of secondary auxiliary views.

- Be able to obtain the true size of an oblique surface.

INTRODUCTION

The principal views (front, top, and side) are sometimes inadequate to fully describe inclined or oblique surfaces. For manufacturing purposes these surfaces must be shown in their true size for dimensioning and visualization. For this reason we need to expand our knowledge beyond orthographic projection. As an example, the inclined surface of Figure 6-1 is not adequately described in the top view because the hole is an ellipse and the surface is not true size. Although the top view is a true projection, it is confusing.

The primary reason for using auxiliary views is to describe the shape of an object whenever it is not adequately described by the three principal orthographic views. Specifically, auxiliary views are used to show the true length of oblique lines; the true distance between points, lines, or planes; the true size, shape, and slope of surfaces; and true angles between lines or planes. Our study will be limited to the true length of oblique lines, the true size of inclined surfaces, and the true size of oblique surfaces. The other uses are beyond the scope of this book and are better left to a descriptive geometry textbook. Such constructions are not well suited for sketching.

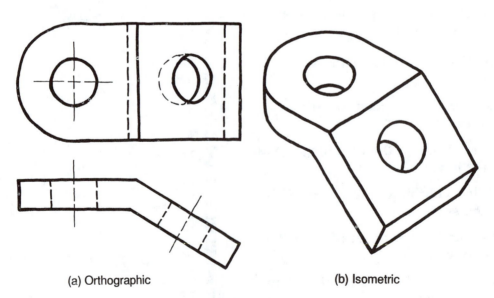

(a) Orthographic

(b) Isometric

FIGURE 6-1
Inadequacy of principal views.

PRINCIPLE OF PRIMARY AUXILIARY VIEWS

A view obtained by projection on any plane except the six principal planes is known as an **auxiliary view.** A **primary auxiliary view** is projected onto a plane that is perpendicular to one of the principal planes and inclined to the other two principal planes. The observer is at an infinite distance, and the parallel lines of sight are perpendicular to the auxiliary plane. Based on this description, an infinite number of primary auxiliary views can be obtained. The location of the auxiliary plane is based on the purpose of the auxiliary view.

Figures 6-2, 6-3, and 6-4 show the auxiliary plane as it relates to the principal planes. In Figure 6-2, pictorial and orthographic views are shown. Both the auxiliary plane and the top plane are perpendicular to the front plane; therefore, we have a folding line T/F between the front view and the top view and another one A/F between the front view and the auxiliary view. This is the basis for the construction of the auxiliary view. When an auxiliary view is taken off the front view, distance d from folding line T/F to point 1 in the top view is the same as distance d from folding line A/F to point 1 in the auxiliary view. Another way of saying the

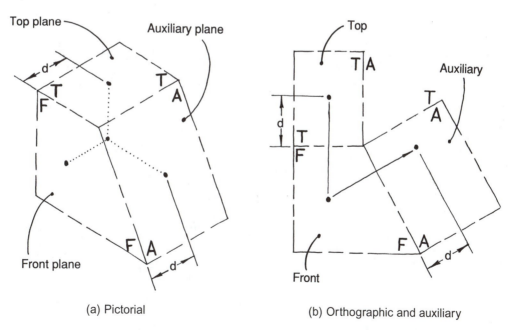

(a) Pictorial

(b) Orthographic and auxiliary

FIGURE 6-2
Auxiliary plane from the front view.

same thing is, "when looking at the top view, folding line T/F represents the edge view of the front projection plane, and when looking at the auxiliary view, the folding line A/F also represents the edge view of the front projection plane; therefore, distances from these folding lines in their respective views must be equal." Projections from one view to an adjacent view cross perpendicular to the folding line.

Figure 6-3 shows the situation where the auxiliary view is taken off the top view. Since both the front projection plane and the auxiliary projection plane are perpendicular to the top projection plane, distances from folding lines T/F and A/T in the front and auxiliary views must be equal. This fact is illustrated by the distance h to point 1 in the front and auxiliary views. Figure 6-4 shows the situation where the auxiliary view is taken off the side view. Since both the top projection plane and the auxiliary projection plane are perpendicular to the side projection plane, distances from folding lines T/S and A/S in the top and auxiliary views must be equal. This fact is illustrated by the distance w to point 1 in the top and auxiliary views. Note that folding line A/T occurs twice in a manner analogous to folding line T/S.

To further illustrate the concept, an object is placed inside the projection box, as shown in Figure 6-5. The purpose of the auxiliary view is to

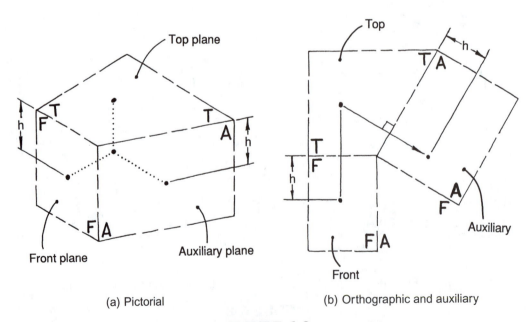

(a) Pictorial

(b) Orthographic and auxiliary

FIGURE 6-3
Auxiliary plane from the top view.

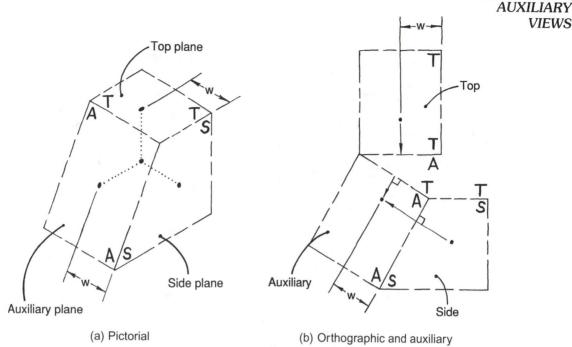

(a) Pictorial

(b) Orthographic and auxiliary

FIGURE 6-4
Auxiliary plane from the side view.

show the true size of the inclined surface. To do this, an auxiliary plane is located parallel to the inclined surface and perpendicular to the front projection plane. The distance d is the distance of the object back of the front plane as shown in the pic-

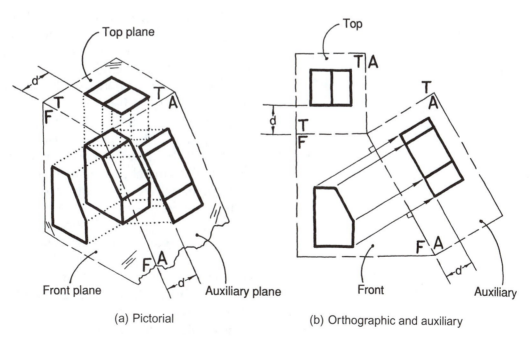

(a) Pictorial

(b) Orthographic and auxiliary

FIGURE 6-5
Auxiliary view from the front view.

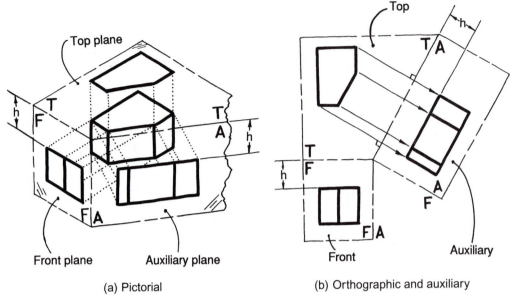

(a) Pictorial   (b) Orthographic and auxiliary

FIGURE 6-6
Auxiliary view from the top view.

torial, the auxiliary view, and the top view. All points, lines, and surfaces must be the same distance from folding lines T/F and A/F in the top and auxiliary views, respectively. The projectors, as illustrated by the thin lines with arrows, are perpendicular to folding line A/F. The association of the orthographic, auxiliary, and pictorial at this time is very helpful.

Figure 6-6 shows another inclined surface; however, the auxiliary view is taken off the top view. The distance h is the same in the front view as in the auxiliary view because both are measured as a vertical distance below the top plane. All points, lines, and surfaces in the front and auxiliary views must be the same distance from the top projection plane. The projectors, as illustrated by the thin lines with arrows, are perpendicular to folding lines A/T. Remember that folding lines T/F and A/T are the edge views of the top projection plane in their respective views.

Similarly, Figure 6-7 shows an inclined surface that requires the auxiliary view to be taken off the side view. The distance w is the same in the top view as in the auxiliary view because both are measured as a horizontal distance to the left of the side plane. All points, lines, and surfaces in the top and auxiliary views must be the same distance from the side projection plane. The projectors, as illustrated

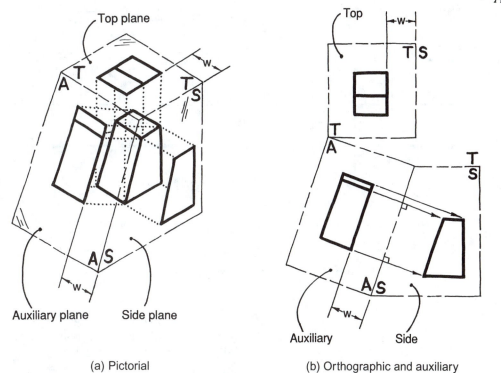

(a) Pictorial

(b) Orthographic and auxiliary

FIGURE 6-7
Auxiliary view from the side view.

by the thin lines with arrows, are perpendicular to folding line A/S. Remember that folding lines T/S and A/S are edge views of the side projection plane in their respective views.

TRUE LENGTH OF A LINE

Many technical problems require the finding of the true length of a line. The true length of normal and inclined lines is given in one of the principal views, as illustrated by Figure 6-8. The information was given in Chapter 3 and is presented at this time as a review. However, the true length of an oblique line must be found by an auxiliary view, as shown in Figure 6-9. In step I an auxiliary plane is constructed parallel to one view of the line. It does not make any difference whether the top, front, or side view is used. In this case the top view is used. The edge view of the top plane as seen in the auxiliary view is designated by the folding line labeled A/T. In step II the points 1 and 2 are projected perpendicular to that folding line. Step III consists of measuring the respective distances from the

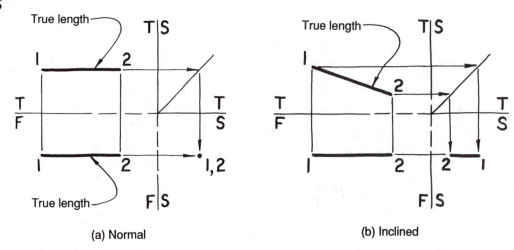

(a) Normal   (b) Inclined

FIGURE 6-8
True length of normal and inclined lines.

points 1 and 2 to the folding line T/F. These distances, *a* and *b*, are transferred to the auxiliary view to locate the points 1 and 2. The line joining these points is the true length. Remember that the folding lines A/T and A/F represent the edge view of the top plane in the auxiliary and front views, respectively.

Note that the side view was not used; however, it could have been because the distances *a* and *b* are given in the side view, too. Figure 6-10

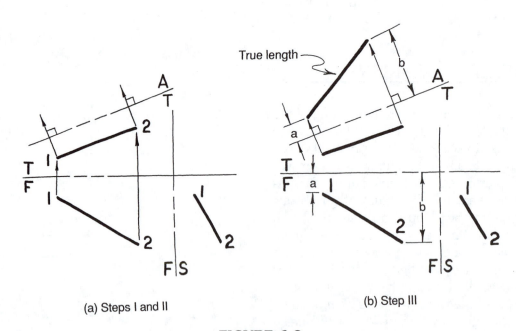

(a) Steps I and II   (b) Step III

FIGURE 6-9
True length of an oblique line
from the top view.

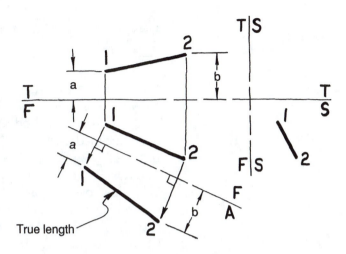

FIGURE 6-10
True length of an oblique line
from the front view.

shows the construction of the true length of a line when the auxiliary is taken off the front view. Note that the distances *a* and *b* can be obtained from either the top or side view. Figure 6-11 shows the construction of the true length of a line when the auxiliary is taken off the side view. Note that the distances *a* and *b* can be obtained from either the top or front view.

In Figures 6-9, 6-10, and 6-11 we are doing the same problem three different ways. If done carefully the measured true length in each case will be the same.

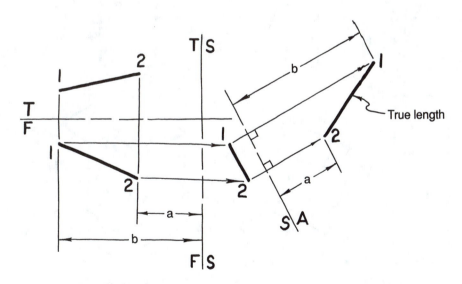

FIGURE 6-11
True length of an oblique line
from the side view.

TRUE SIZE OF AN INCLINED SURFACE

Another important use of auxiliary views is to obtain the true size of an inclined surface. Figure 6-12 shows the method of obtaining the true size of an inclined surface. The method is similar to that used to obtain the true length of a line. In step I the auxiliary plane, as represented by folding line A/F, is established parallel to the edge view of the inclined surface, which is labeled by the numerals 1, 2, 3, and 4. In step II projectors of the points 1,2 and 3,4 are constructed perpendicular to the folding line A/F. The distances *a* and *b* are used to locate the points on the projectors (step III). Lines are constructed between the points as shown at step IV and the remaining points of the object are located in the same manner, thereby giving the completed auxiliary view. The true size of the inclined surface is shown in the auxiliary view. An isometric projection is given at (e) for reference purposes only.

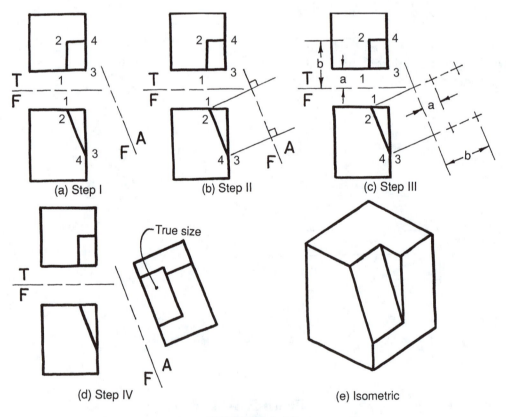

FIGURE 6-12
True size of an inclined surface
from the front view.

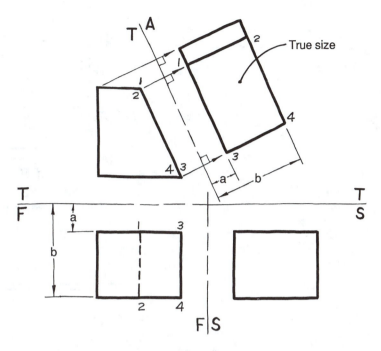

FIGURE 6-13
True size of an inclined surface
from the top view.

Figure 6-13 shows the construction of the true size of an inclined surface taken off the top view. Similarly, Figure 6-14 shows the construction of the true size of an inclined surface taken off the side view.

It is not always necessary to show the complete auxiliary view. In Figure 6-15 only a **partial auxiliary** view is needed because the center of

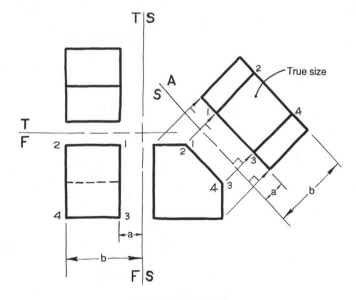

FIGURE 6-14
True size of an inclined surface
from the side view.

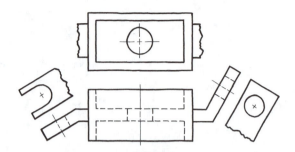

FIGURE 6-15
Partial auxiliary views
of inclined surfaces.

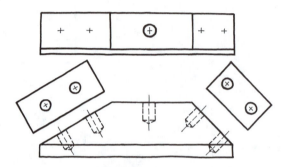

FIGURE 6-16
Partial auxiliary views
of inclined surfaces.

the block is adequately described in the top and front views. The inclined surfaces are shown in detail in the two auxiliary views. In most cases we do not attempt to add the true projection in the top view because this would be time spent needlessly.

Figure 6-16 illustrates the true size of inclined surfaces for dimensioning and visualization. An object of this type would not require all the detail in the top view of the holes on the inclined surfaces. Dimensioning and location of the holes on the inclined surfaces would be done in the auxiliary views. Unless needed for clarity, it is not necessary to construct the complete view of all the inclined surfaces. In fact, a complete auxiliary view can be confusing. Partial auxiliary views are common practice.

TRUE SIZE OF AN OBLIQUE SURFACE

The problem of finding the true size of an oblique surface, the surface 1,2,3 in Figure 6-17, is somewhat more difficult. An oblique surface does not

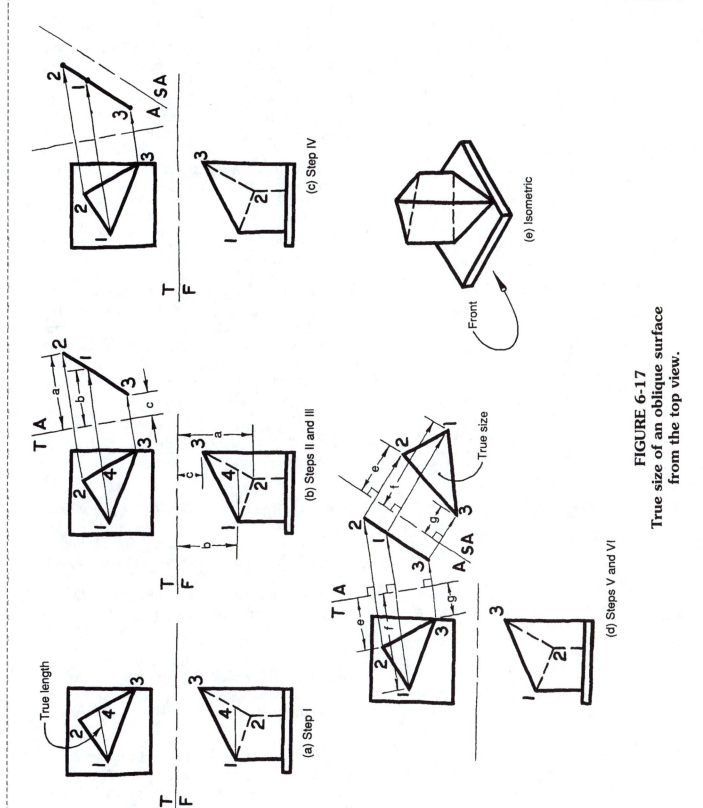

FIGURE 6-17
True size of an oblique surface
from the top view.

show as an edge in any of the three principal views. Therefore, we will need to take two auxiliary views to show an oblique surface in true size. This additional auxiliary view, called a **secondary auxiliary view,** is projected from a primary auxiliary view on a projection plane that is inclined to all the principal projection planes. An infinite number of secondary auxiliary planes is possible; however, only one will achieve the desired result.

Figure 6-17 shows the method of obtaining the true size of an oblique surface. The steps are as follows:

Step I. Obtain the true length of a line on the oblique surface in one of the principal views. A horizontal line 1,4 is constructed on the oblique surface of the front view. Since the line is parallel to the top projection plane (folding line T/F), the view of the line in the top view 1,4 is the true length. Note that in some cases such as Exercise 6-4 the true length of the line is part of the surface and does not need to be constructed.

Step II. An auxiliary plane as presented by folding line T/A is established perpendicular to the true length line 1,4 in the top view.

Step III. The auxiliary view of the oblique surface is a straight line; therefore, we are seeing the edge view of the oblique surface. The method of obtaining this view is the same as previously described under inclined surfaces. Distances of *a, b,* and *c* from folding line T/A are equal to the distance of *a, b* and *c* from folding line T/F and represent the distances from the top projection plane.

Step IV. The secondary auxiliary plane is established parallel to the edge view of the oblique surface and is represented by folding line A/SA.

Step V. Projectors of points 1, 2, and 3 are constructed perpendicular to folding line A/SA. Distances from folding line A/SA to points 1, 2, and 3 are equal distances from folding line T/A to points 1, 2, 3 and are represented by the letters *e, f,* and *g.*

Step VI. The points 1, 2, and 3 are joined to give the true size of the oblique surface. Usually, the rest of the object is not completed because it is adequately represented in the other views.

In this case the auxiliary views were projected from the top view. The same results could have been obtained by projecting from the front or side views and these cases are left as exercises.

Sketching auxiliary views requires great care. The reader may find it desirable to use a small triangle and scale.

SUMMARY

In summary, one must know or do the following:

1. Auxiliary views describe the shape of an object when the principal views are inadequate.

2. An auxiliary plane is located in the desired position and the observer is at an infinite distance.

3. Projectors are perpendicular to the auxiliary plane, and one of the principal planes is seen as an edge.

4. Distances are measured in a previous view showing the edge of the same plane as the auxiliary view.

5. The true length of an oblique line is obtained by an auxiliary plane parallel to any view of the oblique line.

6. The true size of an inclined surface is obtained by an auxiliary plane parallel to the edge view of that surface.

7. Partial auxiliary views are a common practice.

8. The true size of an oblique surface requires a primary auxiliary view and a secondary auxiliary view.

9. The first auxiliary for an oblique surface produces an edge view of that surface.

DIMENSIONING

CHAPTER

7

Upon completion of this chapter, the student is expected to:

- Know the conventions of good dimensioning practice.

- Be able to dimension a three-view sketch.

INTRODUCTION

Our previous discussions in this textbook have centered on the shape or contour of the object as described by the various views (shape description) of
an object. In this chapter and Chapter 8, we will be
concerned with the size of the object and with the
location and form of geometric features of the object. Dimensions are used to supply operating personnel with the necessary information (size, form,
and location) to manufacture the object. They are
not necessarily the same dimensions used to create
the shape description. The object must be fully dimensioned so that operating personnel are not required to add, subtract, or multiply values or to use
a scale to determine a measurement.

When applying dimensions, one should remember the following important criteria:

1. Accuracy: the values are correct.

2. Clearness: most appropriate position.

3. Completeness: no omissions.

4. Readability: lines, values, and notes are legible.

The information in this chapter is based on dimensioning practices as defined by ASME Y14.5M
1994.

NOMENCLATURE

A **dimension line** is a thin line with a break near
its center for insertion of the dimension value. It is
terminated at both ends with an arrowhead whose
length, 3 mm (0.12 inch), is approximately three
times its width, 1 mm (0.04 inch). The arrowhead
touches an **extension line,** which indicates the
relationship of the dimension line to the object. Extension lines begin approximately 1.5 mm (0.06
inch) from the object line and extend approximately 3 mm (0.12 inch) beyond the arrowhead.
Figure 7-1(a) illustrates the correct method of constructing these lines. The lines are thin, and are
usually sketched with a 3H-grade pencil. The correct method of constructing the arrowhead is
shown at (b).

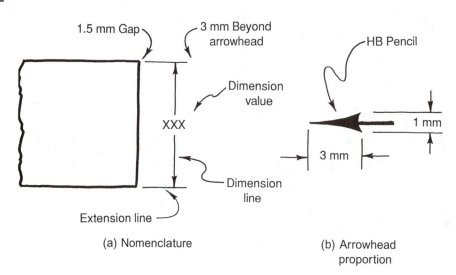

(a) Nomenclature

(b) Arrowhead proportion

FIGURE 7-1
Dimension and extension lines.

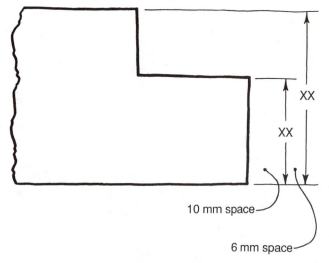

FIGURE 7-2
Space between dimensions.

Figure 7-2 illustrates the space between dimensions. As a general rule, the first dimension line is placed 10 mm (0.40 inch) from the object and all succeeding dimensions are placed 6 mm (0.24 inch) apart.

Because of congested spaces, special techniques are necessary to enable the dimensions to be readable. Figure 7-3(a) illustrates the special technique for extension lines, and at *b* the techniques for dimensioning small values are shown.

Except in special cases, object lines or hidden lines should not be used as extension lines; however, as shown in Figure 7-4(a), centerlines are frequently used as extension lines. Extension lines can cross other extension lines, object lines, hidden lines, or centerlines. In these cases, no gaps occur in the extension line at the crossing point except where there is an arrowhead or an arrowhead close to a dimension line. It is not considered good practice for an extension line to cross a dimension line. Proper ordering of the dimensions from small to large can avoid that situation.

Remember that the object and hidden lines must dominate so that the object stands out. It is not considered good practice to have dimension lines and values within the views of the object as shown at Figure 7-4(a). Dimensioning to hidden lines is to be avoided; however, this is not always possible (see right-side view of Figure 7-6). These conventions cannot be followed at all times; therefore, judgment is usually required.

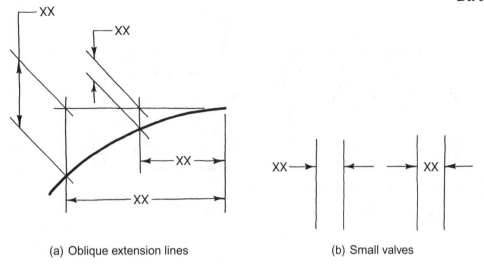

(a) Oblique extension lines   (b) Small valves

FIGURE 7-3
Dimensioning in crowded spaces.

Another dimensioning technique is the use of **leaders.** They are used to direct a dimension, symbol, or local note to its intended place on the drawing as illustrated in Figure 7-5. **Local notes,** such as FULL KNURL, apply to a specific operation and are placed as near as possible to the detail concerned. A straight line is sketched from the local note to the part feature with an arrowhead at the end. For circles, the straight line if extended would pass through the center of the hole. Local notes are printed horizontally. **General notes** differ from local notes in that they apply to the entire ob-

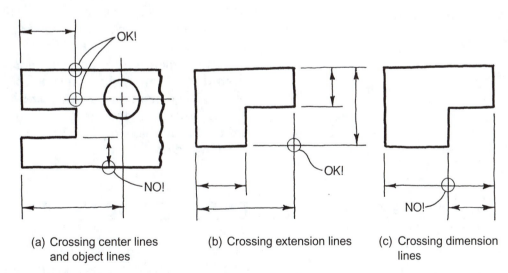

(a) Crossing center lines   (b) Crossing extension lines   (c) Crossing dimension
and object lines   lines

FIGURE 7-4
Crossing other lines.

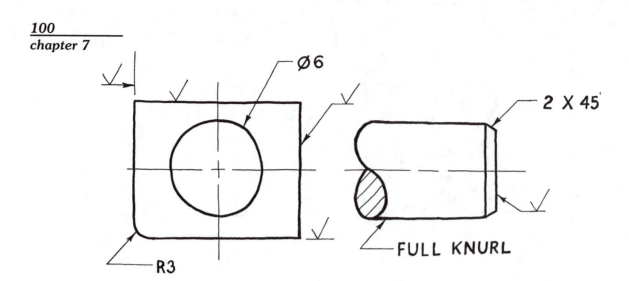

(a) Leaders and finish marks   (b) Local note

FIGURE 7-5
Leaders, finish marks, and local notes.

ject. A typical general note would be ALL DIMEN-SIONS ± 1 mm.

Finish marks are placed on all views of an object where the surface appears as an *edge*, as shown in Figure 7-5. The finish mark is V-shaped with an included angle of 60° and with the right side twice as high (approximately 7 to 8 mm) as the left side. It points toward the bottom of the page. Details such as the roughness height, waviness height, and lay of the surface quality may be incorporated into the symbol. These details are discussed in Chapter 8. Finish marks are omitted on features where a note specifies a machine operation. They should also be omitted on holes and surfaces of finished stock. Dimensioning should be to finished surfaces.

All dimension values and notes are read from the bottom of the drawing. This method of placing the dimensions is called **unidirectional.** The aligned method whereby the values and notes are read from the bottom and the right side is not recommended.

DIMENSIONING VALUES

At one time workers scaled the various views of an object to obtain dimensions. As the manufacture of products became more complicated, the previously described system of dimensioning was developed.

Initially, all dimension values were in whole numbers and fractions of an inch. However, as more precise measurements were needed, the use of fractions was replaced by the decimal inch and the millimeter systems. Fractions will continue to be used for certain commercial commodities that use standardized nominal designations, such as pipe, lumber, and fastener sizes.

When using millimeter dimensioning, the following practices are observed:

1. Where the value is less than 1 millimeter, a zero precedes the decimal point (0.3).

2. Where the value is a whole number, neither the decimal point nor a zero is shown (3).

3. The last digit to the right of the decimal point is not followed by zeros (3.3, not 3.300) unless the value is part of a tolerance specification.

4. Neither commas nor spaces are used to separate groups of digits such as thousands.

When using decimal inch dimensioning, the following practices are observed:

1. Where the value is less than 1 inch, a zero is not used before the decimal point (.25).

2. Zeros are added to the right of the decimal point to indicate accuracy (3.300, not 3.3).

Dual dimensioning whereby both millimeters and inches are given is sometimes used; however, this practice is discouraged. Where a few inch dimensions are shown on a metric drawing, they are followed by IN. Where a few millimeter dimensions are shown on an English drawing, they are followed by mm.

Unless otherwise noted, all dimensions in this book are in millimeters.

DIMENSIONING FOR SIZE

To adequately dimension an object for size, the width, height, and depth of all geometric features should be specified. As a general convention (see Figure 7-6), dimensions for the height of the object should be placed between the views; the width should be placed above the front view; and the depth above the side view or to the right of the top view. As the object becomes more complicated than the one in the figure, this convention cannot be followed in all situations.

Another convention is to place dimensions adjacent to the view that shows the profile or contour of the geometric feature. This is illustrated by the dimensioning of the slot in the top view of Figure 7-6. Unless unavoidable, dimensioning should not be to a hidden line. The location of the slot is not given—dimensioning for location is discussed in the next section.

Whenever possible, related dimensions should be grouped. Figure 7-7(a) illustrates one method of grouping dimensions. In this method there is a functional relationship of surface B and C to A. Figure 7-7(b) illustrates one method of grouping dimensions. In this method there is a functional relationship of surface B to A and surface C to B. The function of the object plays a critical role in determining the method to use.

Figure 7-7(a) also indicates a common error. When dimensions are in series, there is no need for the last dimension since the overall width of the object is given. The convention is to omit the least important dimension in the series. It could, however,

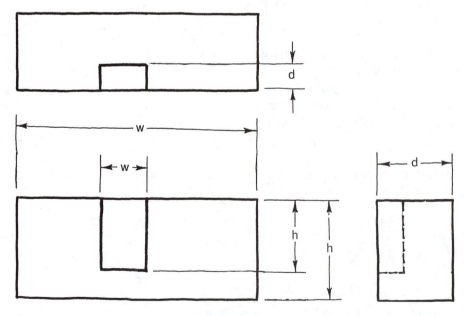

FIGURE 7-6
Placing dimensions.

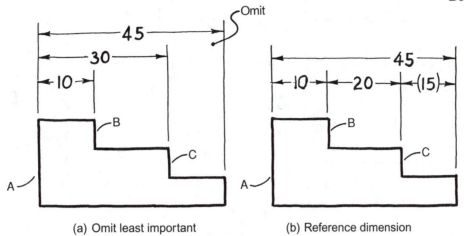

(a) Omit least important   (b) Reference dimension

FIGURE 7-7
Grouping dimensions.

be included as a reference, in which case the value
is enclosed in parentheses.

Miscellaneous shapes have different methods
of dimensioning, as illustrated in Figure 7-8. The
most common shape is the cylinder, and it is dimensioned in the rectangular view as shown in
Figure 7-8(a). At (b) is a sphere with radius 9 mm.
A cone and its method of dimensioning is shown
at (c). A chamfer is dimensioned by two linear dimensions or by a linear and an angular dimension

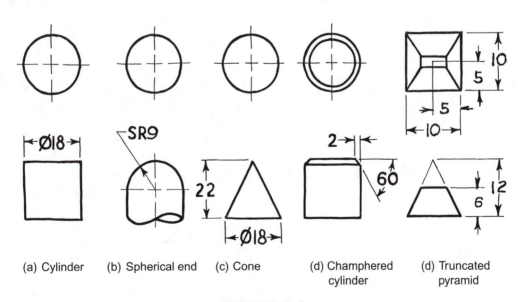

(a) Cylinder   (b) Spherical end   (c) Cone   (d) Champhered cylinder   (d) Truncated pyramid

FIGURE 7-8
Shapes.

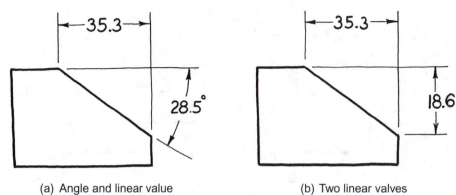

(a) Angle and linear value   (b) Two linear valves

FIGURE 7-9
Angles.

as shown at (d). When the angle is 45°, the chamfer may be dimensioned as shown in Figure 7-5b. The method of dimensioning a pyramid is shown at (e).

Angles of inclined surfaces utilize two methods of dimensioning, depending on the accuracy desired. The least accurate is shown in Figure 7-9a. At (a) the angle is specified by a linear value and an angular one expressed in either decimal degrees as shown or by degrees, minutes, and seconds. The coordinate method as shown at (b) is more accurate because the angle variation (when given in degrees) increases with the distance from the beginning point. Note that the dimension line for the angle is curved.

The dimensioning of curved surfaces has some specific conventions. Figure 7-10 illustrates the method of dimensioning fillets (top) and rounds (bottom). These shapes occur in various processes such as casting, welding, and milling. The letter R precedes the dimensioned value. If there are a number of fillets and rounds with the same radii, a local note such as 3× or a general note may be more appropriate than dimensioning each one. A typical note is ALL ROUNDS AND FILETS R4.

Figure 7-10 can also be used to illustrate the general rules for fillets and rounds as well as some of the rules for all radii. Where location of the center is important and space permits, a dimension line is sketched from the radius center with the arrowhead touching the arc, and the dimension is placed between the arrowhead and the center as shown on the left. If space is limited, the dimen-

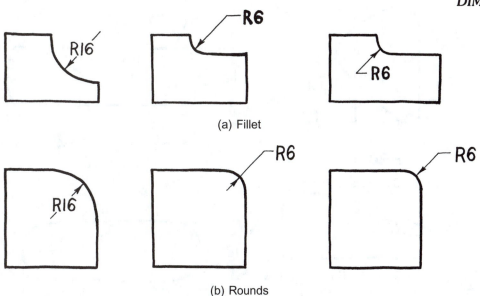

(a) Fillet

(b) Rounds

FIGURE 7-10
Dimensioning of radii.

sion line is extended through the radius center as shown in the middle. Where it is inconvenient to place the arrowhead between the radius center and the arc, it may be placed outside the arc with a leader, as shown on the right. If the center of the radius is not dimensionally located, it is not indicated. In Figure 7-10 none of the centers is dimensionally located—they are controlled by other features, such as tangent surfaces.

Where a dimension is given to the center of a radius, by means of extension and dimension lines, a small cross is sketched at the center as illustrated by Figure 7-11(a). If the dimension line is too long for the space available, it is foreshortened as shown at (c). The method of dimensioning partially rounded ends is shown at (b) and fully rounded ends at (d).

Figure 7-12 illustrates another method of dimensioning round end parts. The holes should be located between centers, and the radius of the end of the part given. Because operating personnel work with centers to locate arcs and holes, the overall dimensions are not normally given. Cylinders are dimensioned in their rectangular view, but holes are usually dimensioned in the circular view by leaders. The depth of the hole is usually given in the local note using the symbol ▼ *after* the value

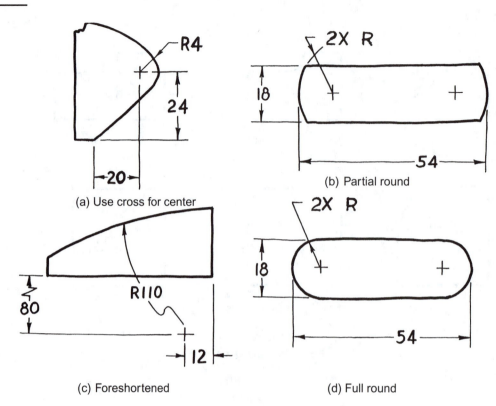

(a) Use cross for center

(b) Partial round

(c) Foreshortened

(d) Full round

FIGURE 7-11
Dimensionally located radius center.

rather than in another view. This enables operating personnel to have the size and depth at the same place. The overall width of the object is 77 mm, and the height is 32 mm. If these values are desired, they are included as references. There are other methods of dimensioning holes and the reader is referred to the ASME standard.

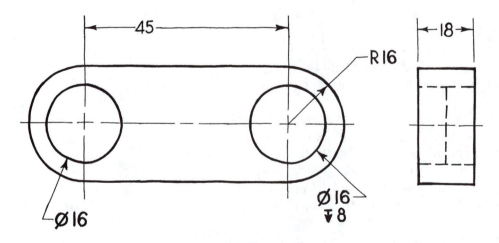

FIGURE 7-12
Round end parts.

A very common geometric feature of an object is a hole for a bolt or screw. Figure 7-13 illustrates the methods of dimensioning these typical fastener holes. The callout note for a counterbored hole is illustrated at (a). First is the diameter symbol, which is followed by the hole size; next is the counterbore symbol ⌴, which is followed by its size; and last is the depth symbol and its value. Counterboring a hole is for the purpose of placing the head of the screw below the surface of the part. It is used largely for socket and fillister head screws. Additional information on fasteners is given in Chapter 9.

The callout note for a countersunk hole is illustrated in Figure 7-13(b). After the diameter symbol and the hole size, the letters THRU are given. This symbol stands for "through" and is used wherever it is not obvious that the hole goes through. Next is the countersunk symbol (∨), which is followed by its diameter and included angle. The countersunk symbol is a wide V with an included angle of 90°. An × between the diameter and the angle is read as "by" and has a space before and after it. Therefore, × can be read as "times" (given previously) or "by." It is important to remember that the diameter is given and not the

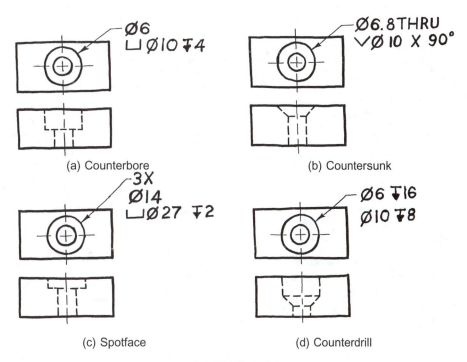

(a) Counterbore

(b) Countersunk

(c) Spotface

(d) Counterdrill

FIGURE 7-13
Typical holes for fasteners.

depth of the countersunk hole, as this is a very difficult and inaccurate measurement to make. Countersunk holes are used with flathead screws to place the head of the screw even with the surface of the part. Because the bearing surface of the head is at an angle, the countersunk hole may be used on thin parts, such as door hangers and thin covers. A special dimensioning technique is used for countersunk holes of curved surfaces.

The method of specifying a spotfacing operation is shown in Figure 7-13(c). In this case there are three identical holes, as given by the 3× in the callout. The × is read as "times" and there is no space between the 3 and the ×. After the diameter of the hole, the spotface symbol (⊔) and its diameter are given. The last part of the symbol is the depth of the spotface. Note that the spotface symbol and the counterbore symbol are the same. The two are quite similar except that spotfacing has a larger diameter and its depth is only a few millimeters. If the depth is not given, the spotface operation is to the minimum necessary to clean up the surface to the specified diameter; or, the depth of the remaining material may be specified in another view. Spotfacing is used to remove the burr left by the drilling operation and/or to provide a bearing surface for a washer or bolt head.

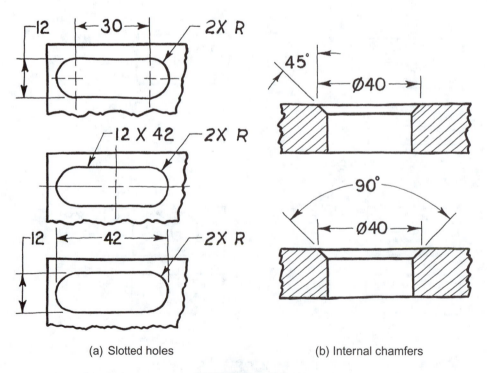

(a) Slotted holes   (b) Internal chamfers

FIGURE 7-14
Slotted holes for internal chambers.

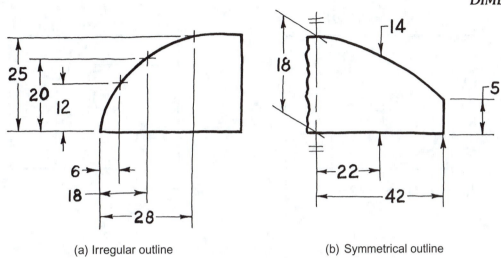

(a) Irregular outline   (b) Symmetrical outline

FIGURE 7-15
Irregular and symmetrical outlines.

Counterdrilled holes, as shown in Figure 7-13(d), are similar to counterbored holes except that the bottom surface has the same shape as a drill bit, which is usually at an included angle of 120°. It is not necessary to specify the included angle.

Slotted holes and internal chamfers present additional dimensioning situations and these are shown in Figure 7-14. Three methods are used for slotted holes. Internal chamfers are dimensioned as shown at (b), with the special case shown by the 45° angle and the general case shown by the 90° angle. If dimensional control is required, two linear dimensions are used.

Irregular outlines are dimensioned as shown in Figure 7-15(a) or the x-y coordinates of the points can be tabulated. At (b) is shown the method of dimensioning symmetrical outlines. In this case the right side is shown and the left side is omitted. Note the break line and the symmetry symbol (=).

DIMENSIONING FOR LOCATION

After the geometric features that describe the shape of an object have been dimensioned, the location of these features is given. In some cases the size dimension will also double as the location dimension. However, in many situations, especially for cylindrical surfaces, another dimension is necessary.

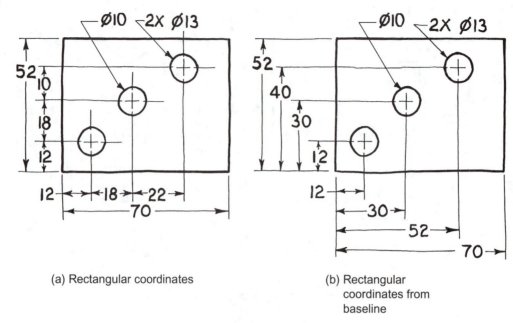

(a) Rectangular coordinates

(b) Rectangular
coordinates from
baseline

FIGURE 7-16
Hole location.

Figure 7-16(a) illustrates one method of indicating the location dimensions for a hole. In this case, the hole on the left is located from two finished surfaces, the middle hole is located from the left hole, and the hole on the right is located from the middle hole. Dimensions are given to the center of the hole in both directions. This illustrates only the location of the holes and not the completely dimensioned part. Avoid dimensioning from unfinished surfaces, because unfinished surfaces cannot be relied upon for accurate measurements. Holes are finished surfaces, even though the finish mark is not used, and their centerlines can be used as extension lines.

Another method is illustrated at Figure 7-16(b). It is referred to as rectangular coordinate dimensioning, and all dimensions are from common baselines which join at the lower left corner of the figure. The method to use is based on the function of the part; however, the latter method is recommended. Cylindrical surfaces should be located in the view in which they appear in their circular form.

Another example of rectangular coordinate dimensioning is shown in Figure 7-17(a). In this case we have a cylindrical object and the coordinates are from the center of the object rather than the lower left corner as shown in Figure 7-16(b). Rectangular coordinate dimensioning where the values are shown on the extension lines and the dimension lines are not used is shown in Figure 7-17(b). The

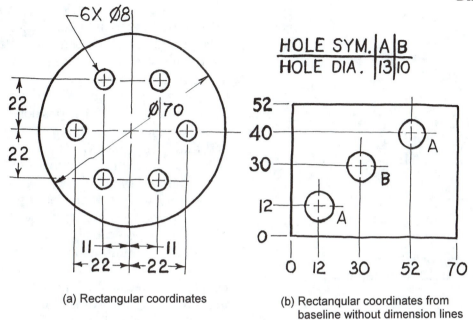

(a) Rectangular coordinates

(b) Rectangular coordinates from baseline without dimension lines

FIGURE 7-17
Rectangular coordinate dimensioning.

baselines are indicated as zero coordinates. Note the method of dimensioning the holes.

Figure 7-18 shows rectangular coordinate dimensioning in tabular form. Note the use of X, Y, and Z to denote the coordinates. Hole size and quantity are also shown in tabular form. Also, the depth of the hole is given under the Z column of the table.

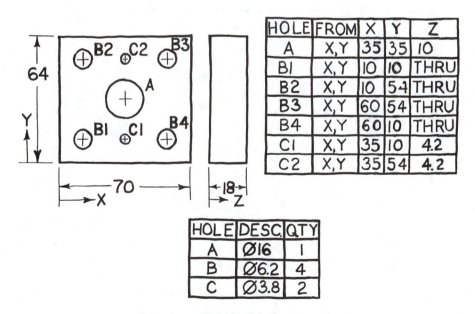

HOLE	FROM	X	Y	Z
A	X,Y	35	35	10
B1	X,Y	10	10	THRU
B2	X,Y	10	54	THRU
B3	X,Y	60	54	THRU
B4	X,Y	60	10	THRU
C1	X,Y	35	10	4.2
C2	X,Y	35	54	4.2

HOLE	DESC.	QTY
A	Ø16	1
B	Ø6.2	4
C	Ø3.8	2

FIGURE 7-18
Rectangular coordinate dimensioning, tabular form.

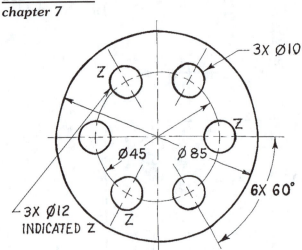

FIGURE 7-19
Bolt circle dimensioning.

The location dimensions for holes on a bolt circle are illustrated in Figure 7-19. The size of the hole, the number of holes, and the spacing are placed in notes. In this case we have 6 holes 60° apart, as indicated by the (6 × 60°) note. If the holes are unequally spaced, an angular dimension can be used to show their position. The bolt circle is always dimensioned in the view where the circle and the holes appear in circular form. This is the view that operating personnel use to lay off and scribe the center line of the holes. If the accuracy of the hole location is critical, a system of coordinate dimensions (similar to those shown in Figure 7-17(a) is used on each hole.

Proper dimensioning of a radial groove is illustrated in Figure 7-20. The groove is produced by an 18-mm-diameter milling cutter; therefore, the width of the groove is given as the diameter of the cutter. Radii at the ends of the groove are not necessary as these are the radius of the cutter. The center radius of the groove (the path of travel of the cutter) and angles locating the end extremities of the cutter are dimensional. To give additional dimensions merely makes it more difficult to machine and harder to check for accuracy. The figure further illustrates the method of dimensioning by polar coordinates, where the radii at the ends of the groove are dimensioned by their angular values (30° and 75°) and a linear value (114).

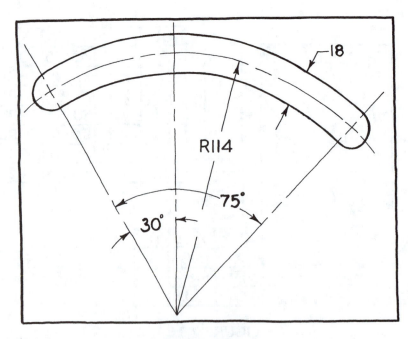

FIGURE 7-20
Radial groove dimensioning.

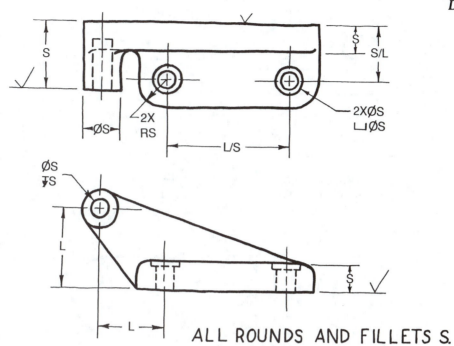

ALL ROUNDS AND FILLETS S.

FIGURE 7-21
Size and location dimensions.

An example of a detailed part showing size and location dimensions is given in Figure 7-21. It should be studied to appreciate the difference between size, S, and location, L, dimensions. Location dimensions locate a part feature, and size dimensions define the physical size of the part feature. In some cases a dimension may be both a size and a location.

ADDITIONAL COMMENTS

While it is important that the views of the object be fully dimensioned, it is equally important that superfluous dimensions be eliminated. Extra dimensions can confuse operating personnel. Figure 7-22 illustrates some superfluous dimensions.

Another important consideration is the function of the part. Formerly, the manufacturing process was considered first and the function second. The present method is to dimension tentatively for the functional aspects, and then review the processes to see if any improvement can be made without adversely affecting the functional relationship.

The dimensioning conventions given in this chapter should be used as a guide. Some judgment

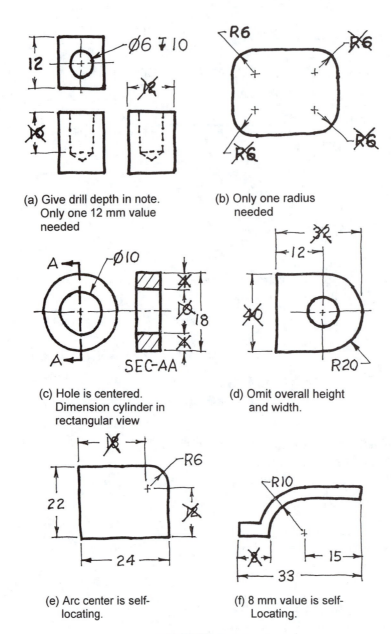

(a) Give drill depth in note. Only one 12 mm value needed

(b) Only one radius needed

(c) Hole is centered. Dimension cylinder in rectangular view

(d) Omit overall height and width.

(e) Arc center is self-locating.

(f) 8 mm value is self-Locating.

FIGURE 7-22
Dimensioning example.

is usually necessary when dimensioning the views of an object. A general review of most conventions is illustrated in Figure 7-23. Notice the placement of dimensions in the principal view and in the view where operating personnel will use them. Hole notes are in the circular view of the holes. Long extension lines and crowding of dimensions and notes have been avoided. As shown here, all dimensions and notes are placed in an open, easy-to-read space.

This chapter has included the major conventions. The reader is referred to the standard for in-

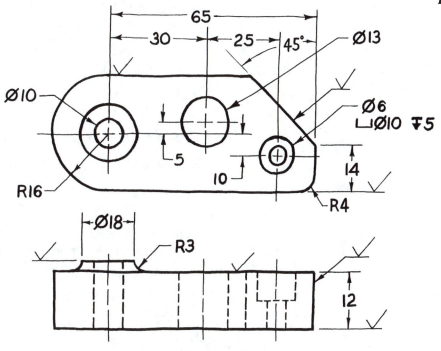

FIGURE 7-23
Dimensioning example.

formation on knurling, keyseats, chords, and limited areas.

SUMMARY

In summary, one must know or do the following:

1. Operating personnel should not have to make calculations or scale a sketch.

2. Dimensioning should be accurate, clear, complete, and readable.

3. The first dimension line should be spaced 10 mm from the view and subsequent ones spaced 6 mm apart.

4. Special techniques are necessary for crowded spaces.

5. Larger dimensions should be outside the smaller dimensions.

6. Visible lines should not be used as extension lines.

7. A centerline may be used as an extension line.

8. Extension lines can cross all types of lines except dimension lines.

9. Visible and hidden lines must stand out from the other lines.

10. Dimensioning should be done off the views of the object when possible.

11. Dimensioning to hidden lines is to be avoided.

12. Leaders are required for local notes.

13. Finish marks are placed on all views of an object where the surface appears as an edge.

14. Finish marks are omitted on features where a note specifies a machining operation, a hole, or on surfaces of finished stock.

15. Dimensions should be given to finished surfaces or hole centers.

16. Dimensions are read from the bottom (unidirectional).

17. In the metric system, a zero precedes a value that is less than 1 (0.3).

18. In the metric system, the last digit to the right of the decimal period is not followed by a zero unless the value is part of a tolerance specification (3.5).

19. In the decimal inch system, a zero does not precede a value that is less than 1 (.5).

20. In the decimal inch system, zeros are added to the right to indicate accuracy (3.500).

21. The width, height, and depth of geometric features should be specified.

22. Cylinders should be dimensioned in their rectangular view.

23. Angles may be dimensioned by degrees and one coordinated dimension or by two coordinate dimensions, depending on the accuracy needed.

24. Reference values are enclosed by parentheses.

25. Holes are dimensioned by leaders.

26. The symbol Ø precedes a diameter value.

27. A small cross is placed at the radius center to indicate dimensional accuracy of the center.

28. When accuracy is desired, the location of holes in a circle is given by coordinate dimensions.

29. A chamfer is dimensioned by two linear values or one linear value and one angular value.

30. Cylindrical surfaces should be located in their circular view.

31. Rectangular coordinate dimensioning is the preferred method for locating holes and similar features.

32. Repetitive linear and angular dimensioning can effectively use the times (×) feature.

33. Polar coordinate dimensioning uses one linear value and one angular value.

34. Symbols are as follows:

S	Sphere
R	Radius
Ø	Diameter
⊔	Counterbore, Spotface
⌄	Countersunk, Counterdrill
√	Finish work
⊤	Depth
×	Times/by
( )	Reference
SØ	Spherical diameter
SR	Spherical radius

35. Superfluous dimensions should be eliminated.

36. In choosing dimensions, function is considered first and processes second.

37. Judgment is necessary to dimension an object.

TOLERANCING

CHAPTER

8

Upon completion of this chapter, the student is expected to:

- Know the definition of tolerance and the techniques of expression.

- Understand the concept of tolerance accumulation.

- Understand the fundamentals of geometric tolerancing.

- Understand tolerance of form, profile, location, orientation, and runout and the techniques of expression.

- Know the nomenclature of mating parts and the three types of fits.

- Know the concept of the shaft basis and hole basis systems and be able to determine the preferred sizes and fits using the metric system.

- Know the concept of surface texture and the techniques of expression.

INTRODUCTION

It is a truism in manufacturing that no two parts can be made exactly alike. There is a certain amount of variation in all the pieces that are produced of a particular part. The sources of this variation are machines, materials, environmental factors, and operating personnel.

The industrial revolution started the practice of interchangeable manufacturing wherein any part could be assembled into any device provided the part was within the prescribed limits. These prescribed limits are determined by the function of the part. For example, a tricycle would not need as tight limits as a watch or jet engine. The tighter the limits are, the more costly the manufacturing operation. Therefore, limits should be specified as loosely as possible provided the function of the device is not compromised.

In this chapter the metric system is used rather than the English system. The concepts are the same—only the numerical values are different. ANSI Y14.5M 1982 is the basis for this chapter.

TOLERANCING PRINCIPLES

Tolerance is the total amount a specific dimension is permitted to vary. It is the difference between the maximum and the minimum limits. In general there are three basic ways of expressing tolerance: direct dimensional tolerancing, geometric tolerancing, and notes.

Geometric tolerancing is discussed in much greater detail in later sections of this chapter. Notes can be quite effective in simplifying the tolerancing effort. A typical note might be UNLESS OTHERWISE SPECIFIED, ALL DIMENSIONS PLUS AND MINUS 0.2. A discussion of direct dimensional tolerancing follows.

Direct dimensional tolerancing takes on two forms: limit dimensioning and plus and minus tolerancing. **Limit dimensioning** is illustrated in Figure 8-1, and the tolerance is expressed by giving the limit dimensions with the high limit placed above the low limit. If the tolerance is expressed on a single line, the low limit is given first and the high limit next with a dash between the two. The same number of decimal places should be used for both

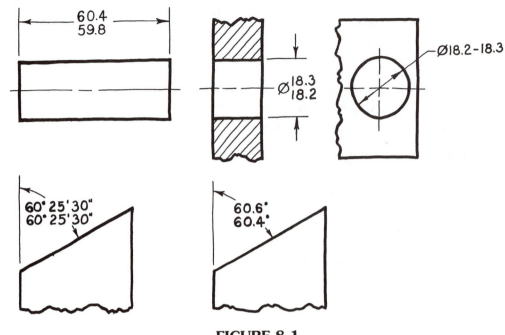

FIGURE 8-1
Limit dimensioning.

limits. Tolerance for angles can be expressed in degrees, minutes, and seconds or by decimal fractions of a degree.

Plus and minus tolerancing is given in Figure 8-2. At (a) the concept of unilateral tolerancing is presented. Only one zero without a plus or minus sign is used. The basic value does not need to have the same number of decimal places as the tolerance. At (b) the concept of bilateral tolerancing is presented. Both the plus and minus values have the same number of decimal places.

All limits regardless of the number of decimal places are used as if they were continued with zeros. For example, 2.3 is equal to 2.300. . .0. A part feature is considered to be in nonconformance if it is outside the limits.

Single limits can be used where the intent is clear, in which case the letters MIN or MAX are placed after the dimension.

When more than one dimension and tolerance occur, a problem is created due to the combining of tolerances. Figure 8-3 shows a comparison of the tolerance accumulation of three different methods of dimensioning. At (a), the **chain dimensioning** method gives the maximum tolerance accumulation between surface b and c of ±0.3 or a total of 0.6 mm. The second

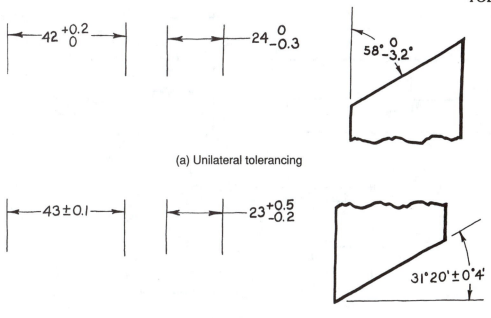

(a) Unilateral tolerancing

(b) Bilateral tolerancing

FIGURE 8-2
Plus and minus tolerancing.

technique is called **baseline dimensioning** and is illustrated in Figure 8-3(b). In this case the tolerance accumulation between surface b and c is ±0.2 or a total of 0.4 mm. The third technique is called **direct dimensioning** and is illustrated in Figure 8-3(c). In this case, the tolerance between surface b and c is 0.1 or a total of 0.2 mm, which is the lowest of the three techniques.

In certain cases it is necessary to indicate that a dimension between two features originates at one of the features and not the other. This technique is shown by a small circle, as at the bottom of Figure 8-3(c).

When there are two or more adjacent dimensions with tolerances, it is important to consider the effect of one tolerance on another in relation to the desired function of the part. The dimensioning technique that is used can be instrumental in preventing an undesirable tolerance accumulation.

Different manufacturing processes are capable of different degrees of accuracy. Table 8-1 illustrates the approximate tolerance range for some shop processes with a basic size of 20 mm. It is also true that the larger the size of the part feature, the larger

TABLE 8-1
Tolerance for Various Processes at 20 mm

PROCESS	TOLERANCE RANGE (MM)
Lapping and honing	0.006 to 0.009
Grinding	0.009 to 0.021
Reaming	0.013 to 0.084
Turning	0.021 to 0.130
Drilling	0.084 to 0.130

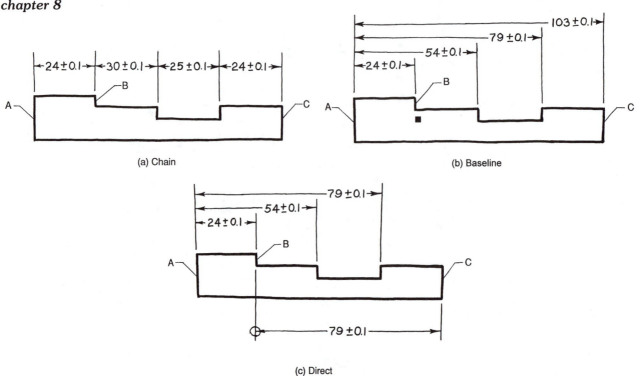

FIGURE 8-3
Tolerance accumulation.

the tolerance. For example, a turning operation for a 100-mm-diameter feature can have a tolerance between 0.035 and 0.220, which is larger than the tolerance given in the table for a 20-mm diameter.

FUNDAMENTALS OF GEOMETRIC TOLERANCING

In simple terms, **geometric tolerancing** is described as a means of specifying the geometry of a part feature. It gives the designer a clear method of expressing the design intent, which in turn enables the manufacturer to choose the most appropriate method to produce the part. In addition, geometric tolerancing determines how the feature is to be inspected and gauged.

A part **feature** is a general term used to describe a physical portion of a part such as a surface, hole, or slot. On the other hand, a **feature of size** is a cylindrical or spherical surface or a set of parallel surfaces that is associated with a size dimension. Size dimensions were discussed in Chapter 7, as were location dimensions, which locate the centerline or centerplane of a part feature with another part feature, centerline, or centerplane.

To simplify the presentation of geometric tolerancing on a sketch, the following techniques are used:

1. Geometric characteristic symbols

2. Datum

3. Basic dimensioning

4. Material condition modifiers

5. Feature control frame

Geometric characteristic symbols are shown in Figure 8-4. Fourteen symbols are used and they are divided into groups. The *tolerance of form* group has the characteristics of straightness, flatness, circularity, and cylindricity. This group is used for individual features. The *tolerance of profile* group has the characteristics of profile of a line and profile of a surface. This group is used with either individual features or related features. The *tolerance of orientation* group has the characteristics of angularity, perpendicularity, and parallelism; the *tolerance of location* group has the characteristics of position, concentricity, and symmetry; and the *tolerance of runout* has the characteristics of circular runout and total runout. These three groups are used with two or more related features. The five groups are discussed in greater detail later in the chapter.

A **datum** indicates the origin of a dimensional relationship between a toleranced feature and designated feature(s) on a part. Letters are used to specify a datum. Figure 8-5 shows two datums, L and K, for a cylindrical surface and a plane surface, respectively. The triangle can be filled or unfilled. It is applied to the surface outline, extension line, dimension line, or feature control frame.

A **basic dimension** is a theoretical exact dimension without tolerance. It is a numerical value that is enclosed in a rectangle, as shown by the dimensions 40.0 and 15.0 in Figure 8-5.

There are three material condition modifiers: maximum material condition (MMC), least material condition (LMC), and regardless of feature size (RFS). These modifiers can be used only with features of size, such as holes, shafts, and slots, rather than with features such as surfaces.

Type of Tolerance	Characteristic	Symbol
Form	Straightness	—
	Flatness	▱
	Circularity	○
	Cylindricity	⌭
Profile	Profile of a Line	⌒
	Profile of a Surface	⌓
Orientation	Angularity	∠
	Perpendicularity	⊥
	Parallelism	∥
Location	Position	⌖
	Concentricity	◎
	Symmetry	≡
Runout	Circular Runout	↗
	Total Runout	⌰

FIGURE 8-4
Geometric characteristic symbols.

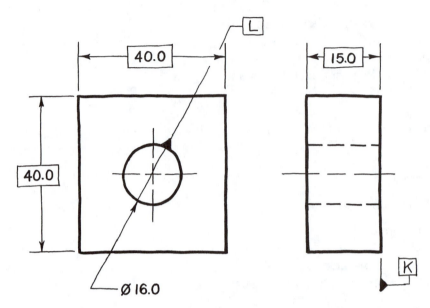

FIGURE 8-5
Datum and basic dimensioning.

When the feature of size contains the greatest amount of material it is at the **maximum material condition.** The MMC of an external feature of size (shaft) is the largest size and the MMC of an internal feature of size (hole) is the smallest size, as shown in Figure 8-6.

When the feature of size contains the least amount of material, it is at the **least material**

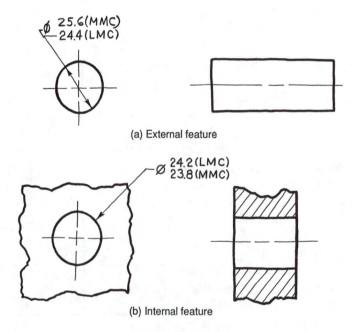

FIGURE 8-6
MMC and LMC of features of size.

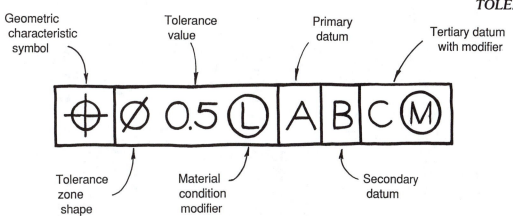

FIGURE 8-7
Typical feature control frame.

condition. The LMC of an external feature of size (shaft) is the smallest size and the LMC of an internal feature of size (hole) is the largest size, as shown in Figure 8-6.

When the feature of size is not at either extreme, but at any size including MMC and LMC, it is referred to as **regardless of feature size.** The RFS occurs when a geometric tolerance is applied independent of the feature size. In other words the geometric tolerance is given as the stated amount for each feature of size.

A **feature control frame** specifies the type, shape, and size of the geometric tolerance zone. It also dictates the datum features and assigns the material condition modifiers to the tolerance and the datum when applicable. A typical feature control frame is shown in Figure 8-7.

TWO RULES OF TOLERANCING

In geometric tolerancing, two fundamental rules form the foundation of the system. **Rule 1** states that where only a tolerance of size is specified, the limits of size prescribe the amount of variation permitted in its form as well as size. Figure 8-8 illustrates this concept for the straightness characteristic, using an external feature at the top and an internal feature at the bottom. The feature of size is shown in Fig. 8-8(a) and its meaning in Fig. 8-8(b). Note that when the actual size is at the MMC value, no variation in form is permitted. In other words, perfect form exists at MMC. Where the actual size

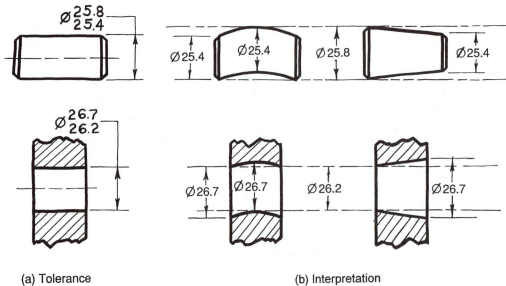

(a) Tolerance

(b) Interpretation

FIGURE 8-8
Straightness example for rule 1.

of the part feature departs from the MMC value, a variation of form is allowed. This variation is equal to the amount of that departure. For a shaft with a tolerance value of Ø 25.4-25.8 the variation in form is as follows:

ACTUAL SIZE	VARIATION IN FORM
25.4 LMC	0.4
25.5	0.3
25.6	0.2
25.7	0.1
25.8 MMC (perfect)	0

For a hole with a tolerance value of Ø 26.2-26.7, the variation in form is as follows:

ACTUAL SIZE	VARIATION IN FORM
26.2 MMC (perfect)	0
26.3	0.1
26.4	0.2
26.5	0.3
26.6	0.4
26.7 LMC	0.5

Note that Figure 8-8 only shows rule 1 for straightness. Similar interpretations for flatness, circularity, and cylindricity are left as exercises. Also, rule 1 does not apply to stock sizes such as bars, sheets, tubing, and so on, or to nonrigid parts such as some plastics. If perfect form at MMC is not required, then it must be so stated.

Rule 2 states that RFS applies with respect to the tolerance, the datum, or both, unless MMC is specified. Furthermore, it is not necessary to include the RFS symbol in the feature control frame. The rule is illustrated in Figure 8-9.

DATUM REFERENCING

Datum planes are associated with process or inspection equipment such as machine tables, surface plates, collets, and locating pins. Although these planes are not true planes, they are of such quality that they are used to simulate the true planes, from which measurements are made to verify dimensions and tolerances. A **datum feature** is the actual feature of a part that is used to establish a datum.

To describe a part accurately, three mutually perpendicular planes are required. These three planes are shown in Figure 8-10 and are called the **datum reference frame.** A part is placed in contact with the primary datum, where three points are needed to make contact with the datum. Two points of contact are needed for the secondary datum, and one point is needed for the tertiary datum. The primary datum plane is listed first, the secondary datum next, and the tertiary datum last in the feature control frame; however, depending on need, only the primary datum or primary and secondary datums may be necessary. Datum features and datum symbols that are related to the datum planes are shown by a two-view sketch in Fig. 8-10(b).

For cylindrical parts, the three-plane reference frame is somewhat more difficult to visualize. The primary datum, D, is the flat surface perpendicular to axis of the cylindrical datum feature as shown in Figure 8-11. Datum E is associated with two theoretical planes at 90° to each other that intersect at the axis of the cylinder. These two planes are represented on the sketch by the centerlines of

$-$	0.5		RFS Applies
$\angle$	0.4	G	RFS Applies
$\angle$	0.3 Ⓜ	F	MMC is specified

FIGURE 8-9
Feature control frame examples of rule 2.

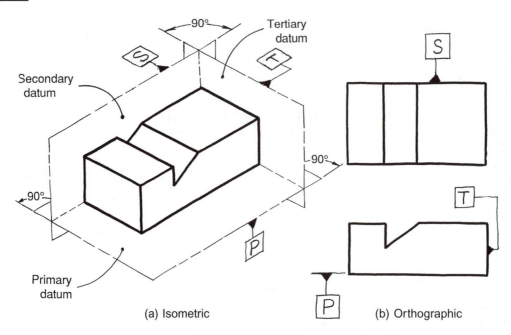

FIGURE 8-10
Datum reference frame.

the cylinder and are used to indicate the direction of measurement in an *x* and *y* direction. Note that the datum symbol is placed with the cylinder limits.

There are two types of datum features: surface and size. A datum surface has no size toler-

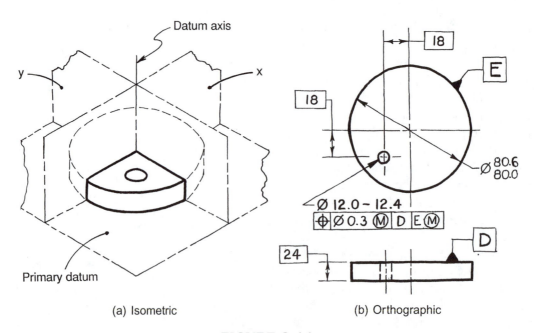

FIGURE 8-11
Cylindrical datum features.

ance because it is a plane from which dimensions or relationships originate. Where a normally flat surface is specified as a datum feature, the corresponding datum is simulated by a plane contacting the high points of that surface. The extent of contact depends on whether the surface is a primary, secondary, or tertiary datum feature.

On the other hand, datums of size are established from features of size with tolerances such as slots and holes. Because variations are allowed by the size dimension, it is necessary to determine whether RFS or MMC will be used. RFS is implied unless otherwise specified (rule 2).

Where a datum feature of size is applied on a RFS basis, the datum is established by physical contact between the feature surface and the surface of the processing equipment. For example, a chuck on a lathe makes contact at different part diameters, thereby establishing a RFS datum for each part.

Where a datum feature of size is applied on a MMC basis, processing and inspection equipment, which remain constant in size, may be used to simulate a true geometric counterpart of the feature and to establish the datum. The size of the datum is the MMC value.

Instead of using a plane to locate a part in a datum reference frame, datum targets can be used. **Datum targets** are designated points, lines, or areas of contact. They are described on the sketch of the part; however, they refer to the gauge features that are used to simulate the datum planes.

A datum target **point** is indicated on the part sketch with an X as shown in Figure 8-12. The point contact is made possible by spherical or pointed locating pins. Datum target symbols are placed outside the part outline with a leader to the target. A solid leader indicates that the target is visible and a dashed leader indicates that the target is invisible, which is the case in the figure. The symbol is a circle with a horizontal line creating an upper and a lower half. A letter and a numeral are placed in the lower half to identify the datum (C) and the point (1) on the datum. Information for the upper half of the circle is given later in this section. The datum target point is usually located by basic dimensioning with gauge tolerances assumed to apply.

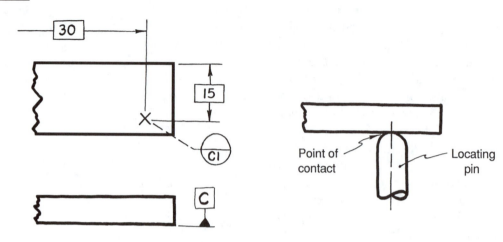

FIGURE 8-12
Datum target point.

Datum target lines are used whenever points are unable to support a part adequately. Also, points might damage a part. However, lines are not as accurate in locating a part as points. An X and a phantom line are used to indicate a target line, as shown in Figure 8-13.

Datum target **areas** are used where additional support for the part is necessary and accuracy is not critical. The area is specified with a crosshatched circle surrounded by a phantom line or an X if a circle is impractical. As shown in Figure 8-14, the diameter of the area is given in the upper

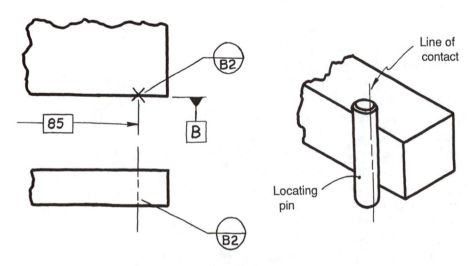

FIGURE 8-13
Datum target line.

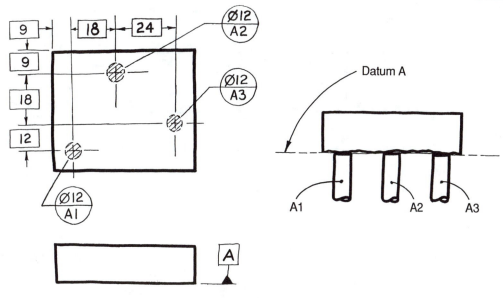

FIGURE 8-14
Datum target area.

half of the symbol. The figure illustrates the use of the areas of three locating pins to simulate the true or theoretical datum plane A.

TOLERANCE OF FORM

Form tolerances control flatness, straightness, circularity, and cylindricity. These tolerances are specified where the tolerances of size or location do not provide sufficient control. They are applicable to individual (single) features and are, therefore, not related to datums.

Flatness is the condition of a surface having all elements in one plane. A flatness tolerance specifies a tolerance zone defined by two parallel planes within which the entire surface must lie, as shown in Figure 8-15. The feature control frame is attached to an extension line of the surface or to a leader with an arrowhead pointing to the applicable surface. It is placed in a view where the surface is represented by a line.

Straightness is the condition in which an *element* of a surface or an axis is a straight line. A straightness tolerance specifies a tolerance zone within which the considered element of the surface or axis of a cylinder must lie. It differs from flatness in that flatness covers an entire surface rather than

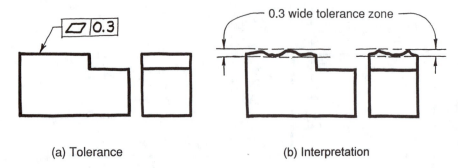

(a) Tolerance   (b) Interpretation

FIGURE 8-15
Flatness tolerance.

an element of the surface. When straightness is specified for a surface, the feature control frame is attached to an extension of a surface element or to a leader with an arrowhead pointing to a surface element. Figure 8-16 illustrates the concept for a cylindrical surface. Each longitudinal element of the circular surface must lie within two parallel lines that are parallel to the axis of the cylinder. The straightness tolerance must be within the limits of size tolerance and the boundary of perfect form at MMC. Figure 8-17 illustrates the concept for a flat surface. Straightness may be applied to line elements in a single direction or to line elements in two directions as shown in the figure. Each longitudinal element must lie between two parallel lines 0.06 apart in the front view and 0.09 in the side view.

Where straightness is specified for an axis, the feature control frame is placed with the feature size

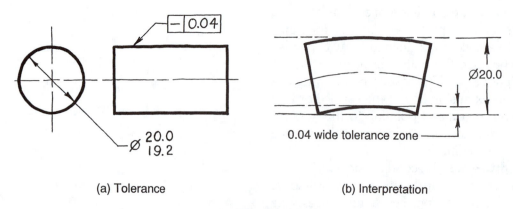

(a) Tolerance   (b) Interpretation

FIGURE 8-16
Straightness tolerance
for a cylindrical surface.

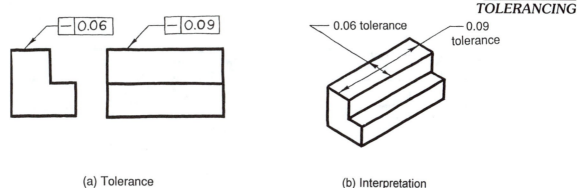

(a) Tolerance   (b) Interpretation

FIGURE 8-17
Straightness tolerance
for a flat surface.

dimension. If a cylindrical feature is being specified, the diameter symbol is required to indicate a cylindrical tolerance zone. The foregoing information is shown in Figure 8-18. It was stated previously that rule 1 controls the four tolerances of form. However, when the axis straightness is applied to a shaft or hole, the feature is no longer required to meet the perfect form at MMC. The absence of a modifier assumes RFS. When the part feature is at 18.2, dimension x can be 18.5, and when the part feature is at 18.8, dimension x can be 19.1, which is the maximum value and is called the **virtual condition.** *Note:* If MMC had been specified, the diameter tolerance zone would have increased as the part feature departed from the MMC value and would equal 0.9 at LMC.

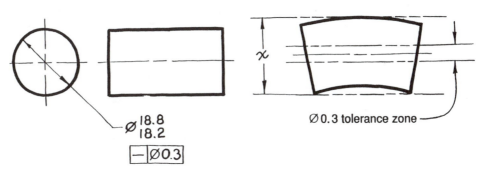

(a) Tolerance   (b) Interpretation

FIGURE 8-18
Straightness tolerance
for an axis.

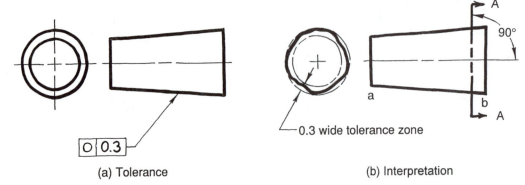

(a) Tolerance

(b) Interpretation

FIGURE 8-19
Circularity tolerance.

Circularity is a condition of a surface of revolution (cylinder, cone, or sphere) where all points of the surface intersected by any plane perpendicular to a common axis (cylinder or cone) or passing through a common center (sphere) are equidistant from the center. A circularity tolerance specifies a tolerance zone bounded by two concentric *circles* within which the circular element must lie. Figure 8-19 illustrates the concept. The tolerance applies independently at any plane along the feature from point *a* to point *b*.

Cylindricity is a condition of a surface of revolution in which all points of the surface are equidistant from a common axis. A cylindricity tolerance specifies a tolerance zone bounded by two concentric *cylinders* within which the surface must lie. The tolerance callout and interpretation are shown in Figure 8-20. Cylindricity tolerance is a combination of circularity and straightness applied to a cylindrical object.

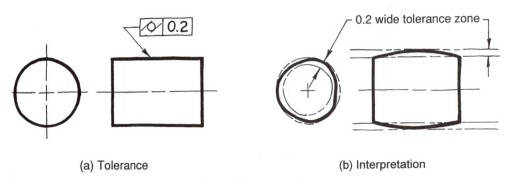

(a) Tolerance

(b) Interpretation

FIGURE 8-20
Cylindricity tolerance.

TOLERANCE OF PROFILE

Profile tolerancing is a method of controlling irregular surfaces, lines, arcs, or unusual shapes, as well as regular shapes. Profiles can be applied to individual line elements called **profile of a line** or to the entire surface called **profile of a surface** of a part. The difference between profile of a line and profile of a surface is analogous to the difference between circularity and cylindricity. Profile tolerancing is unique in that it can be applied to individual features (no datum) or to related features (with a datum).

A profile tolerance specifies a tolerance zone along the true profile within which the elements of the surface or the entire surface must lie. The profile of a surface tolerance is three-dimensional and may be applied to surfaces having a constant cross section, whereas the profile of a line is two-dimensional and applies to parts having a varying cross section, such as the tapered wing of an airplane. The tolerance zone is the distance between two boundaries shaped to the true profile that has been defined by basic dimensions. Two different methods of specifying the tolerance zone can be used as shown in Figure 8-21. In the bilateral method the zone is divided on both sides of the true profile and the boundaries do not need to be shown. However, in the unilateral method a boundary (phantom line) is needed to show the direction of the zone from the true profile.

In most cases a datum reference is necessary to provide proper orientation for a profile of a surface. For a profile of a line, a datum is usually not needed. If it is necessary to indicate the extent of the profile, that information is given by the all-around symbol as shown on the leader in Fig. 8-21(b). It also is shown by the between symbol below the feature control frame in Fig. 8-21(e). It is also possible to specify a different tolerance for different portions of a surface or line. A profile tolerance can be combined with other types of geometric tolerancing.

TOLERANCE OF ORIENTATION

Angularity, perpendicularity, and parallelism are the three orientation tolerances. They control the orientation of a feature or features to a datum feature. All three can be applied to a feature or a fea-

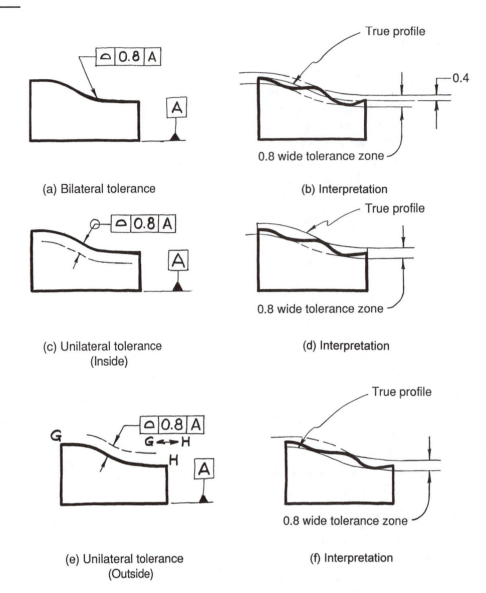

(a) Bilateral tolerance

(b) Interpretation

(c) Unilateral tolerance
(Inside)

(d) Interpretation

(e) Unilateral tolerance
(Outside)

(f) Interpretation

FIGURE 8-21
Profile tolerance of a surface.

ture of size and can use a material condition modifier. When they are applied to plane surfaces, the flatness of the surface is controlled to within the orientation tolerance.

Angularity is the condition of a surface or axis at a specified angle (other than 90°) from a datum plane or axis. An angularity tolerance specifies a tolerance zone defined by two parallel planes at a specific angle with the datum. All points on the angular surface or angular axis must lie between these two planes. Figure 8-22 illustrates the concept for a surface. Note that the angle is enclosed in the basic box that is required for an angularity tolerance.

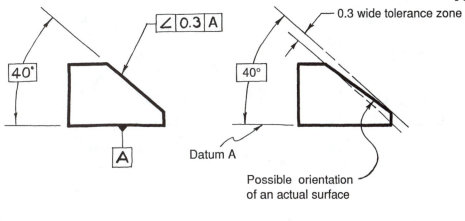

(a) Tolerance

(b) Interpretation

FIGURE 8-22
Angularity tolerance of a surface.

Perpendicularity is the condition of a surface, center plane, or axis at a right angle to a datum plane or axis. A perpendicularity tolerance specifies a tolerance zone defined by two parallel planes, two parallel lines, or a cylinder at 90° to a datum. All elements of the surface, center plane, or axis must fall within the tolerance. Figure 8-23 illustrates the concept for a perpendicular surface.

When the perpendicularity tolerance zone is a cylinder, the diameter symbol must be included in the feature control frame as shown in Figure 8-24. This figure also illustrates the application of the tolerance to a feature of size. In this case the diameter of the cylindrical tolerance zone remains at 0.2

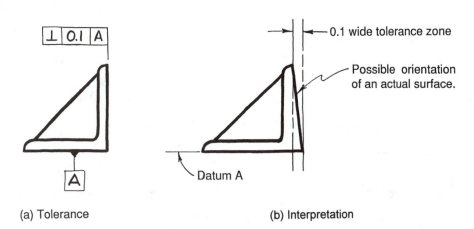

(a) Tolerance

(b) Interpretation

FIGURE 8-23
Perpendicularity tolerance
of a surface.

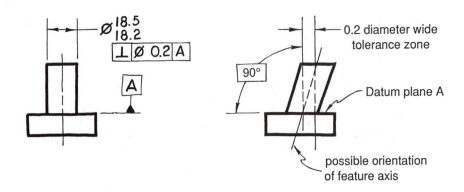

(a) Tolerance   (b) Interpretation

FIGURE 8-24
Perpendicularity tolerance
of an axis.

since RFS is assumed. However, if MMC had been specified after the 0.2 value in the feature control frame, the tolerance zone diameter would vary from 0.2 at MMC to as much as 0.5 at LMC. Thus, as the feature of size departs from the MMC value, a bonus tolerance occurs.

Parallelism is the condition of a surface equidistant at all points from a datum plane or an axis equidistant along its length to a datum axis. A parallelism tolerance specifies a tolerance zone defined by two parallel planes, two parallel lines, or a cylinder that is parallel to a datum. Figure 8-25 illustrates the concept for a surface.

TOLERANCE OF LOCATION

Position, concentricity, and symmetry are the three tolerances of location. Location controls deal with

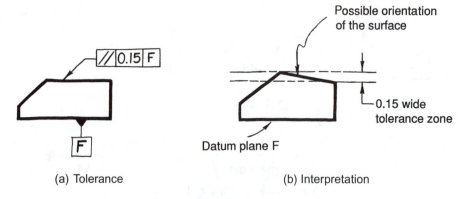

(a) Tolerance   (b) Interpretation

FIGURE 8-25
Parallelism tolerance of surface.

features of size only and require one of the three material condition modifiers to be specified in the feature control frame. Also, locational controls require at least one datum reference.

Positional tolerance defines a zone in which the center, axis, or center plane is permitted to vary from the exact (perfect) location, which is referred to as its *true position*. A positional tolerance is the total permissible variation in the location of the feature about its true position. The positional tolerance is indicated by its symbol, tolerance, modifier, and datum reference in the feature control frame. For cylindrical features, the tolerance is generally the diameter of the tolerance zone that contains the axis (true position) of the feature of size. For features that are not round, such as slots and tabs, the positional tolerance is the total width of the tolerance zone in which the center plane (true position) of the feature must lie.

Figure 8-26(a) shows the tolerance with basic dimensions establishing the true position of the axis of the hole. In Fig. 8-26(b) is shown the interpretation of the tolerance. When limit or plus and minus dimensions are used to locate the axis, a square tolerance zone occurs as shown in Fig. 8-26(c). This latter technique is not recommended because most fixed pin gauges use round pins, which cannot evaluate a square tolerance zone. In addition, tolerance accumulation occurs when a second feature of size is located from the first.

A positional tolerance applied at MMC may be explained in terms of the surface of a hole or in

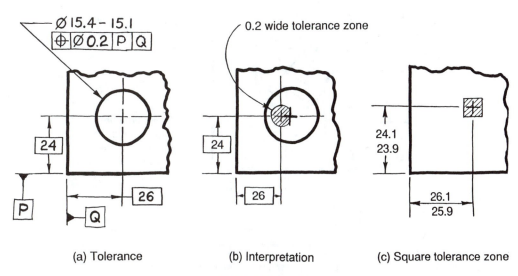

(a) Tolerance　　　(b) Interpretation　　　(c) Square tolerance zone

FIGURE 8-26
Positional tolerance.

terms of the axis of a hole. For a surface, the hole position may vary, but no element of its surface can be inside the theoretical boundary. As shown in Figure 8-27, the theoretical boundary is the MMC minus the positional tolerance.

Where a hole is at MMC, its axis must fall within a cylindrical tolerance zone whose axis is located at the true position. The diameter of this zone is equal to the positional tolerance. Figure 8-28(a) shows the tolerance with its interpretation shown in Fig. 8-28(b). Three holes are used to illustrate the concept. The hole at the left has the axis centered in the cylindrical tolerance zone; the hole in the middle has the axis on the edge of the zone; and the hole at the right has the axis on an angle. In all three cases the axis is inside the tolerance zone. It is only when the feature is at MMC that the specified tolerance applies. As the feature of size departs from the MMC, an additional positional tolerance results. For example, where the feature of size is 15.5, the positional tolerance is $\emptyset$ 0.4. Similarly, for features of size of 15.6 and 15.7, the positional tolerances are $\emptyset$ 0.5 and 0.6, respectively.

Occasionally, there is a need to extend a tolerance zone beyond the surface of the part. This action can be indicated by using the projected tolerance symbol, $\textcircled{P}$, and specifying the height of the

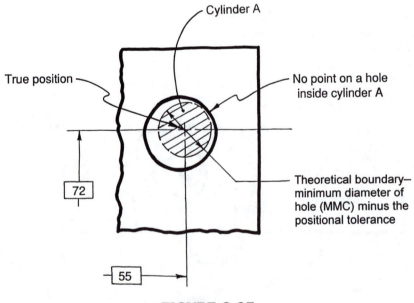

FIGURE 8-27
Boundary for a surface
of a hole at MMC.

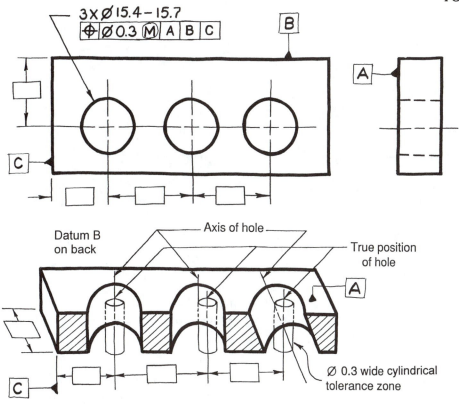

(b) Interpretation

FIGURE 8-28
Hole axis in relation to positional
tolerance zones.

theoretical tolerance zone above the surface. Figure 8-29 illustrates the concept. Note that the symbol and height of the zone are given in a box below the feature control frame.

In certain cases the design or function of a part may require the positional tolerance, datum reference, or both to use an RFS material condition modifier. Since each feature of size axis must be located within the specified tolerance zone, the requirement imposes a closer control of the features. Also, verification of the specification is quite intricate.

Where positional tolerancing at LMC is used, the stated positional tolerance applies to its LMC

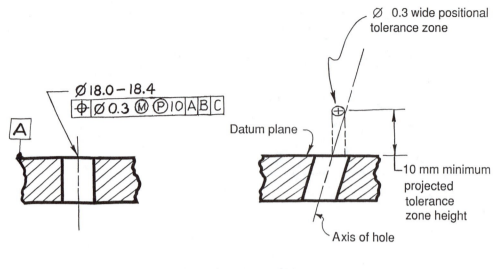

(a) Tolerance

(b) Interpretation

FIGURE 8-29
Projected tolerance zone.

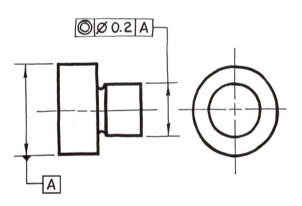

(a) Tolerance

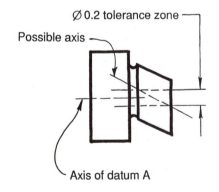

(b) Interpretation

FIGURE 8-30
Concentricity tolerance.

size. Perfect form is required at LMC. Where the feature departs from its stated LMC size, an increase in the positional tolerance is allowed and is equal to the amount of the departure. This additional tolerance is the same as discussed previously under MMC. Specifying LMC is not too common unless considerations critical to the design are required.

Concentricity is the condition where the axis of all cross-sectional elements of a surface of revolution (feature of size) are common to the axis of a datum reference. A concentricity tolerance specifies a cylindrical tolerance zone whose axis coincides with a datum axis (usually the larger diameter) and within which the axis of the considered feature (smaller diameter) must lie. Figure 8-30 illustrates the concept. A concentricity tolerance is established only on an RFS basis. Verification of the tolerance is difficult; therefore, concentricity specifications should be avoided where possible.

Symmetry is a condition in which a feature or features are symmetrically disposed about the center plane of a datum feature. A symmetrical tolerance is specified by two parallel planes within which the center plane of the feature must lie. Figure 8-31 illustrates the concept for a slot. In this case the center plane of the slot must lie between two parallel planes 0.8 apart, regardless of the feature size, which are equally disposed about the center plane of datum B.

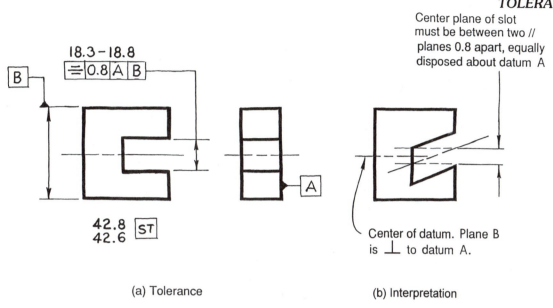

(a) Tolerance

(b) Interpretation

FIGURE 8-31
Symmetry tolerance.

Note the use of the statistical tolerance symbol after the 42.8-42.6 tolerance, which means that the tolerance is based on statistical tolerancing. If it is a statistical geometric tolerance the symbol is placed in the feature control frame following the tolerance and any modifier. If it is a statistical size tolerance, the symbol is placed adjacent to the size dimension.

TOLERANCE OF RUNOUT

Circular runout and total runout are the two tolerances of runout. **Runout** is combination control affecting both the form (surface) and location (position) of a part feature to a datum axis. The types of surfaces controlled by the runout tolerances are those surfaces constructed around the datum axis and those surfaces perpendicular to the datum axis as shown in Figure 8-32. Runout control requires a datum reference and the material condition modifier is RFS. A runout tolerance value is the full indicator movement, FIM, of the considered feature when the part is rotated 360° about its datum axis. Runout is used most frequently on objects that have circular cross sections resulting from turning operations such as would occur on a lathe.

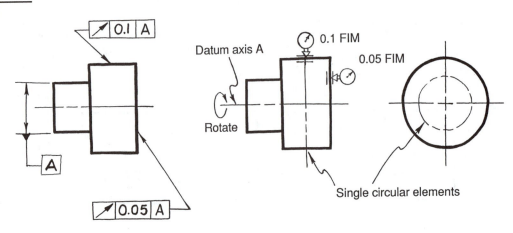

(a) Tolerance

(b) Interpretation

FIGURE 8-32
Circular runout tolerance.

Circular runout provides control of circular *elements* of a surface. Figure 8-32 shows the application of the tolerance to both a surface constructed around a datum axis and one perpendicular to a datum axis. The tolerance is applied independently to any circular element.

Total runout provides combination control of all surface elements. The tolerance is applied simultaneously to circular and longitudinal elements as the part is rotated 360°. Figure 8-33 shows the application of the tolerance to both a surface constructed around a datum axis and one perpendicu-

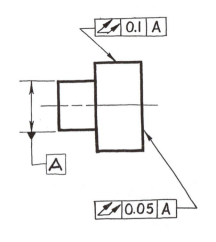

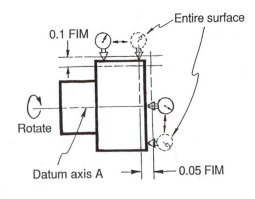

(a) Tolerance

(b) Interpretation

FIGURE 8-33
Total runout tolerance.

lar to a datum axis. Where it is applied to surfaces around the axis, it controls the cumulative variations of circularity, cylindricity, straightness, coaxiality, angularity, taper, and profile. Where it is applied to surfaces perpendicular to the axis, it controls cumulative variations of perpendicularity and flatness.

LIMITS AND FITS FOR MATING PARTS

Whenever two parts are designed to function together, they are mating parts. One of the mating parts is classified as external and the other as internal. The shaft and hole is the most common external-internal pair; however, the reader should note that there are other pairs, such as key-keyway, T–bolt-slot, and electrical plug-outlet. Our discussion will use the shaft-hole pair to illustrate the concept. The basis of the material in this section is ANSI B4.2-1978, Preferred Metric Limits and Fits. A similar system is used for the English system.

To use the system, an understanding of the nomenclature is required. Figure 8-34 shows a

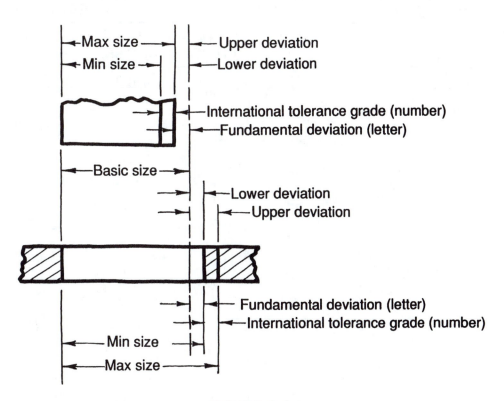

FIGURE 8-34
Nomenclature for mating parts.

shaft-hole pair and the appropriate terms. For simplicity the shaft and the hole are sketched coincident to each other on the left side. The maximum and minimum allowable size of the shaft and of the hole are shown in the figure. As stated previously, the difference in the maximum and minimum values is the tolerance.

The **basic size** is a value from which the deviations are applied to determine the maximum and minimum sizes of the parts. These values should be selected from Table 8-2. Whenever possible, you should select basic sizes from the first column because they represent standard sizes for materials and tools, thereby reducing cost. Also, the tables are based on these preferred sizes.

The **upper deviation** is the difference between the maximum permissible size of a part and the basic size. For a shaft this difference is the largest shaft (MMC) minus the basic size; for a hole this difference is the largest hole (LMC) minus the basic size. The **lower deviation** is the difference

TABLE 8-2
Preferred Sizes

BASIC SIZE (MM)		BASIC SIZE (MM)		BASIC SIZE (MM)	
FIRST CHOICE	SECOND CHOICE	FIRST CHOICE	SECOND CHOICE	FIRST CHOICE	SECOND CHOICE
1		10		100	
	1.1		11		110
1.2		12		120	
	1.4		14		140
1.6		16		160	
	1.8		18			180
2		20		200	
	2.2		22			220
2.5		25		250	
	2.8		28			280
3		30		300	
	3.5		35			350
4		40		400	
	4.5		45			450
5		50		500	
	5.5		55			550
6		60		600	
	7		70			700
8		80		800	
	9		90			900
				1000	

between the minimum permissible size of a part and the basic size. For a shaft, this difference is the smallest shaft (LMC) minus the basic size; for a hole, this difference is the smallest hole (MMC) minus the basic size. The deviation that is closest to the basic size is called the **fundamental deviation.** For the shaft, it is the upper deviation; for the hole, it is the lower deviation and is specified with an uppercase letter such as H. Note that in both cases, the fundamental deviation involves the MMC.

An **international tolerance (IT) grade** is used to specify the tolerance. There are 18 IT grades, ranging from IT01, IT0, IT1, to IT16. Within each grade the tolerances vary in accordance with the basic size. This system provides a uniform level of accuracy.

The **tolerance zone** is a combination of the fundamental deviation and the tolerance grade as shown in Figure 8-35. For a shaft, the basic size is first, then the shaft fundamental deviation (lowercase letter), and then the IT grade; for a hole the basic size is first, then the hole fundamental deviation (uppercase letter), and then the IT grade. The fit is specified by the basic size, hole tolerance zone, and then the shaft tolerance zone. Four methods of showing the tolerance are given in Figure 8-36. Where information is given in parentheses, it is for reference only.

There are three types of fits: clearance, interference, and transition. A **clearance fit** occurs where there is *always* an air space between two

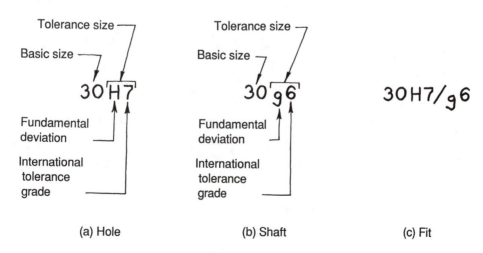

(a) Hole   (b) Shaft   (c) Fit

FIGURE 8-35
Tolerance symbols for mating parts.

$$30\,H7 \qquad 30\,H7\binom{30.021}{30.000} \qquad \begin{matrix}30.021\\30.000\end{matrix} \qquad \begin{matrix}30.021\\30.000\end{matrix}\Bigl(30H7\Bigr)$$

(a)   (b)   (c)   (d)

FIGURE 8-36
Methods for specifying tolerance.

mating parts regardless of the part sizes (provided that they are within tolerance). As can be seen in Figure 8-37(a), the shaft is smaller than the hole under all tolerance conditions. An **interference fit** occurs where there is *always* an interference of material between two mating parts regardless of the part sizes (provided that they are within tolerance). As can be seen in Fig. 8-37(b), the shaft is larger than the hole under all tolerance conditions. The assembly of these two mating parts requires that they be forced together. A **transition fit** is one where *sometimes* there is an air space and *sometimes* there is material interference, depending on mating part sizes. This concept is illustrated in Fig. 8-37(c). When the shaft is at MMC (largest shaft) and the hole is at MMC (smallest hole), there is material interference; however, when both features are at LMC, there is an air space. Transition fits are used with location dimensions.

Preferred fits for clearance, transition, and interference are given in Table 8-3. Wherever possi-

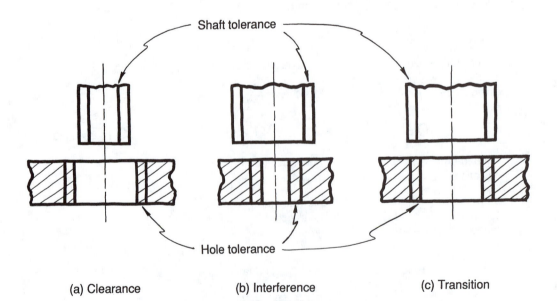

Shaft tolerance

Hole tolerance

(a) Clearance   (b) Interference   (c) Transition

FIGURE 8-37
Hole tolerance.

TABLE 8-3
Description of Preferred Fits

HOLE BASIS	SHAFT BASIS	DESCRIPTION
		Clearance Fits
H11/c11	C11/h11	*Loose-running* fit for wide commercial tolerances or allowances on external members.
H9/d9	D9/h9	*Free-running* fit—not for use where accuracy is essential, but good for large temperature variations, high running speeds, or heavy journal pressures.
H8/f7	F8/h7	*Close-running* fit for running on accurate machines and for accurate location at moderate speeds and journal pressures.
H7/g6	G7/h6	*Sliding* fit—not intended to run freely, but to move and turn freely and locate accurately.
		Transition Fits
H7/h6	H7/h6	*Locational clearance* fit provides snug fit for locating stationary parts, but can be freely assembled and disassembled.
H7/k6	K7/h6	*Locational transition* fit for accurate location, a compromise between clearance and interference.
H7/n6	N7/h6	*Locational transition* fit for more accurate location where greater interference is permissible.
		Interference Fits
H7/p6[a]	P7/h6	*Locational interference* fit for parts requiring rigidity and alignment with prime accuracy of location but without special bore pressure requirements.
H7/s6	S7/h6	*Medium drive* fit for ordinary steel parts or shrink fits on light sections, the tightest fit usable with cast iron.
H7/u6	U7/h6	*Force* fit suitable for parts that can be highly stressed or for shrink fits where the heavy pressing forces required are impractical.

[a]Transition fit for basic sizes in range from 0 through 3 mm.

ble, fits should be selected from this table. The description provides information for the designer to select the appropriate fit. Fits are given for hole basis and shaft basis. These two systems were developed to aid the practitioner to determine the tolerances of mating parts.

The **hole basis system** uses the smallest hole (MMC) as the basic size; therefore, its lower deviation is 0. It is the most common system because standard reamers, broaches, and other tools are used to produce holes and standard plug gages are used to inspect them. Shafting, on the other hand, can be machined to any desired size.

A practical example will aid in understanding the system. The designer finds that a free-running fit will meet the criteria for two mating parts. Also, design calculations give a diameter of approximately 29 mm.

TABLE 8-4
Preferred Hole Basis Fits for Diameters to 30 mm

BASIC SIZE		LOOSE RUNNING			FREE RUNNING			CLOSE RUNNING			SLIDING			LOCATIONAL CLEARANCE		
		HOLE H11	SHAFT C11	FIT	HOLE H9	SHAFT D9	FIT	HOLE H8	SHAFT F7	FIT	HOLE H7	SHAFT G6	FIT	HOLE H7	SHAFT H6	FIT
1	MAX	1.060	0.940	0.180	1.025	0.980	0.070	1.014	0.994	0.030	1.010	0.998	0.018	1.010	1.000	0.016
	MIN	1.000	0.880	0.060	1.000	0.955	0.020	1.000	0.984	0.006	1.000	0.992	0.002	1.000	0.994	0.000
1.2	MAX	1.260	1.140	0.180	1.225	1.180	0.070	1.214	1.194	0.030	1.210	1.198	0.018	1.210	1.200	0.016
	MIN	1.200	1.080	0.060	1.200	1.155	0.020	1.200	1.184	0.006	1.200	1.192	0.002	1.200	1.194	0.000
1.6	MAX	1.660	1.540	0.180	1.625	1.580	0.070	1.614	1.594	0.030	1.610	1.598	0.018	1.610	1.600	0.016
	MIN	1.600	1.480	0.060	1.600	1.555	0.020	1.600	1.584	0.006	1.600	1.592	0.002	1.600	1.594	0.000
2	MAX	2.060	1.940	0.180	2.025	1.980	0.070	2.014	1.994	0.030	2.010	1.998	0.018	2.010	2.000	0.016
	MIN	2.000	1.880	0.060	2.000	1.955	0.020	2.000	1.984	0.006	2.000	1.992	0.002	2.000	1.994	0.000
2.5	MAX	2.560	2.440	0.180	2.525	2.480	0.070	2.514	2.494	0.030	2.510	2.498	0.018	2.510	2.500	0.016
	MIN	2.500	2.380	0.060	2.500	2.455	0.020	2.500	2.484	0.006	2.500	2.492	0.002	2.500	2.494	0.000
3	MAX	3.060	2.940	0.180	3.025	2.980	0.070	3.014	2.994	0.030	3.010	2.998	0.018	3.010	3.000	0.016
	MIN	3.000	2.880	0.060	3.000	2.955	0.020	3.000	2.984	0.006	3.000	2.992	0.002	3.000	2.994	0.000
4	MAX	4.075	3.930	0.220	4.030	3.970	0.090	4.018	3.990	0.040	4.012	3.996	0.024	4.012	4.000	0.020
	MIN	4.000	3.855	0.070	4.000	3.940	0.030	4.000	3.978	0.010	4.000	3.988	0.004	4.000	3.992	0.000
5	MAX	5.075	4.930	0.220	5.030	4.970	0.090	5.018	4.990	0.040	5.012	4.996	0.024	5.012	5.000	0.020
	MIN	5.000	4.855	0.070	5.000	4.940	0.030	5.000	4.978	0.010	5.000	4.988	0.004	5.000	4.992	0.000
6	MAX	6.075	5.930	0.220	6.030	5.970	0.090	6.018	5.990	0.040	6.012	5.996	0.024	6.012	6.000	0.020
	MIN	6.000	5.855	0.070	6.000	5.940	0.030	6.000	5.978	0.010	6.000	5.988	0.004	6.000	5.992	0.000
8	MAX	8.090	7.920	0.260	8.036	7.960	0.112	8.022	7.987	0.050	8.015	7.995	0.029	8.015	8.000	0.024
	MIN	8.000	7.830	0.080	8.000	7.924	0.040	8.000	7.972	0.013	8.000	7.986	0.005	8.000	7.991	0.000
10	MAX	10.090	9.920	0.260	10.036	9.960	0.112	10.022	9.987	0.050	10.015	9.995	0.029	10.015	10.000	0.024
	MIN	10.000	9.830	0.080	10.000	9.924	0.040	10.000	9.972	0.013	10.000	9.986	0.005	10.000	9.991	0.000
12	MAX	12.110	11.905	0.315	12.043	11.950	0.136	12.027	11.984	0.061	12.018	11.994	0.035	12.018	12.000	0.029
	MIN	12.000	11.795	0.095	12.000	11.907	0.050	12.000	11.966	0.016	12.000	11.983	0.006	12.000	11.989	0.000
16	MAX	16.110	15.905	0.315	16.043	15.950	0.136	16.027	15.984	0.061	16.018	15.994	0.035	16.018	16.000	0.029
	MIN	16.000	15.795	0.095	16.000	15.907	0.050	16.000	15.966	0.016	16.000	15.983	0.006	16.000	15.989	0.000
20	MAX	20.130	19.890	0.370	20.052	19.935	0.169	20.033	19.980	0.074	20.021	19.993	0.041	20.021	20.000	0.034
	MIN	20.000	19.760	0.110	20.000	19.883	0.065	20.000	19.959	0.020	20.000	19.980	0.007	20.000	19.987	0.000

Size																
25	MAX	25.130	24.890	0.370	25.052	24.935	0.169	25.033	24.980	0.074	25.021	24.993	0.041	25.021	25.000	0.034
	MIN	25.000	24.760	0.110	25.000	24.883	0.065	25.000	24.959	0.020	25.000	24.980	0.007	25.000	24.987	0.000
30	MAX	30.130	29.890	0.370	30.052	29.935	0.169	30.033	29.980	0.074	30.021	29.993	0.041	30.021	30.000	0.034
	MIN	30.000	29.760	0.110	30.000	29.883	0.065	30.000	29.959	0.020	30.000	29.980	0.007	30.000	29.987	0.000
1	MAX	1.010	1.006	0.010	1.010	1.010	0.006	1.010	1.012	0.004	1.010	1.020	-0.004	1.010	1.024	-0.008
	MIN	1.000	1.000	-0.006	1.000	1.004	-0.010	1.000	1.006	-0.012	1.000	1.014	-0.020	1.000	1.018	-0.024
1.2	MAX	1.210	1.206	0.010	1.210	1.210	0.006	1.210	1.212	0.004	1.210	1.220	-0.004	1.210	1.224	-0.008
	MIN	1.200	1.200	-0.006	1.200	1.204	-0.010	1.200	1.206	-0.012	1.200	1.214	-0.020	1.200	1.218	-0.024
1.6	MAX	1.610	1.606	0.010	1.610	1.610	0.006	1.610	1.612	0.004	1.610	1.620	-0.004	1.610	1.624	-0.008
	MIN	1.600	1.600	-0.006	1.600	1.604	-0.010	1.600	1.606	-0.012	1.600	1.614	-0.020	1.600	1.618	-0.024
2	MAX	2.010	2.006	0.010	2.010	2.010	0.006	2.010	2.012	0.004	2.010	2.020	-0.004	2.010	2.024	-0.008
	MIN	2.000	2.000	-0.006	2.000	2.004	-0.010	2.000	2.006	-0.012	2.000	2.014	-0.020	2.000	2.018	-0.024
2.5	MAX	2.510	2.506	0.010	2.510	2.510	0.006	2.510	2.512	0.004	2.510	2.520	-0.004	2.510	2.524	-0.008
	MIN	2.500	2.500	-0.006	2.500	2.504	-0.010	2.500	2.506	-0.012	2.500	2.514	-0.020	2.500	2.518	-0.024
3	MAX	3.010	3.006	0.010	3.010	3.010	0.006	3.010	3.012	0.004	3.010	3.020	-0.004	3.010	3.024	-0.008
	MIN	3.000	3.000	-0.006	3.000	3.004	-0.010	3.000	3.006	-0.012	3.000	3.014	-0.020	3.000	3.018	-0.024
4	MAX	4.012	4.009	0.011	4.012	4.016	0.004	4.012	4.020	0.000	4.012	4.027	-0.007	4.012	4.031	-0.011
	MIN	4.000	4.001	-0.009	4.000	4.008	-0.016	4.000	4.012	-0.020	4.000	4.019	-0.027	4.000	4.023	-0.031
5	MAX	5.012	5.009	0.011	5.012	5.016	0.004	5.012	5.020	0.000	5.012	5.027	-0.007	5.012	5.031	-0.011
	MIN	5.000	5.001	-0.009	5.000	5.008	-0.016	5.000	5.012	-0.020	5.000	5.019	-0.027	5.000	5.023	-0.031
6	MAX	6.012	6.009	0.011	6.012	6.016	0.004	6.012	6.020	0.000	6.012	6.027	-0.007	6.012	6.031	-0.011
	MIN	6.000	6.001	-0.009	6.000	6.008	-0.016	6.000	6.012	-0.020	6.000	6.019	-0.027	6.000	6.023	-0.031
8	MAX	8.015	8.010	0.014	8.015	8.019	0.005	8.015	8.024	0.000	8.015	8.032	-0.008	8.015	8.037	-0.013
	MIN	8.000	8.001	-0.010	8.000	8.010	-0.019	8.000	8.015	-0.024	8.000	8.023	-0.032	8.000	8.028	-0.037
10	MAX	10.015	10.010	0.014	10.015	10.019	0.005	10.015	10.024	0.000	10.015	10.032	-0.008	10.015	10.037	-0.013
	MIN	10.000	10.001	-0.010	10.000	10.010	-0.019	10.000	10.015	-0.024	10.000	10.023	-0.032	10.000	10.028	-0.037
12	MAX	12.018	12.012	0.017	12.018	12.023	0.006	12.018	12.029	0.000	12.018	12.039	-0.010	12.018	12.044	-0.015
	MIN	12.000	12.001	-0.012	12.000	12.012	-0.023	12.000	12.018	-0.029	12.000	12.028	-0.039	12.000	12.033	-0.044
16	MAX	16.018	16.012	0.017	16.018	16.023	0.006	16.018	16.029	0.000	16.018	16.039	-0.010	16.018	16.044	-0.015
	MIN	16.000	16.001	-0.012	16.000	16.012	-0.023	16.000	16.018	-0.029	16.000	16.028	-0.039	16.000	16.033	-0.044
20	MAX	20.021	20.015	0.019	20.021	20.028	0.006	20.021	20.035	-0.001	20.021	20.048	-0.014	20.021	20.054	-0.020
	MIN	20.000	20.002	-0.015	20.000	20.015	-0.028	20.000	20.022	-0.035	20.000	20.035	-0.048	20.000	20.041	-0.054
25	MAX	25.021	25.015	0.019	25.021	25.028	0.006	25.021	25.035	-0.001	25.021	25.048	-0.014	25.021	25.061	-0.027
	MIN	25.000	25.002	-0.015	25.000	25.015	-0.028	25.000	25.022	-0.035	25.000	25.035	-0.048	25.000	25.048	-0.061
30	MAX	30.021	30.015	0.019	30.021	30.028	0.006	30.021	30.035	-0.001	30.021	30.048	-0.014	30.021	30.061	-0.027
	MIN	30.000	30.002	-0.015	30.000	30.015	-0.028	30.000	30.022	-0.035	30.000	30.035	-0.048	30.000	30.048	-0.061

TABLE 8-5
Preferred Shaft Basis Fits for Diameters to 30 mm

BASIC SIZE		LOOSE RUNNING			FREE RUNNING			CLOSE RUNNING			SLIDING			LOCATIONAL CLEARANCE		
		HOLE H11	SHAFT C11	FIT	HOLE H9	SHAFT D9	FIT	HOLE H8	SHAFT F7	FIT	HOLE H7	SHAFT G6	FIT	HOLE H7	SHAFT H6	FIT
1	MAX	1.120	1.000	0.180	1.045	1.000	0.070	1.020	1.000	0.030	1.012	1.000	0.018	1.010	1.000	0.016
	MIN	1.060	0.940	0.060	1.020	0.975	0.020	1.006	0.990	0.006	1.002	0.994	0.002	1.000	0.994	0.000
1.2	MAX	1.320	1.200	0.180	1.245	1.200	0.070	1.220	1.200	0.030	1.212	1.200	0.018	1.210	1.200	0.016
	MIN	1.260	1.140	0.060	1.220	1.175	0.020	1.206	1.190	0.006	1.202	1.194	0.002	1.200	1.194	0.000
1.6	MAX	1.720	1.600	0.180	1.645	1.600	0.070	1.620	1.600	0.030	1.612	1.600	0.018	1.610	1.600	0.016
	MIN	1.660	1.540	0.060	1.620	1.575	0.020	1.606	1.590	0.006	1.602	1.594	0.002	1.600	1.594	0.000
2	MAX	2.120	2.000	0.180	2.045	2.000	0.070	2.020	2.000	0.030	2.012	2.000	0.018	2.010	2.000	0.016
	MIN	2.060	1.940	0.060	2.020	1.975	0.020	2.006	1.990	0.006	2.002	1.994	0.002	2.000	1.994	0.000
2.5	MAX	2.620	2.500	0.180	2.545	2.500	0.070	2.520	2.500	0.030	2.512	2.500	0.018	2.510	2.500	0.016
	MIN	2.560	2.440	0.060	2.520	2.475	0.020	2.506	2.490	0.006	2.502	2.494	0.002	2.500	2.494	0.000
3	MAX	3.120	3.000	0.180	3.045	3.000	0.070	3.020	3.000	0.030	3.012	3.000	0.018	3.010	3.000	0.016
	MIN	3.060	2.940	0.060	3.020	2.975	0.020	3.006	2.990	0.006	3.002	2.994	0.002	3.000	2.994	0.000
4	MAX	4.145	4.000	0.220	4.060	4.000	0.090	4.028	4.000	0.040	4.016	4.000	0.024	4.012	4.000	0.020
	MIN	4.070	3.925	0.070	4.030	3.970	0.030	4.010	3.988	0.010	4.004	3.992	0.004	4.000	3.992	0.000
5	MAX	5.145	5.000	0.220	5.060	5.000	0.090	5.028	5.000	0.040	5.016	5.000	0.024	5.012	5.000	0.020
	MIN	5.070	4.925	0.070	5.030	4.970	0.030	5.010	4.988	0.010	5.004	4.992	0.004	5.000	4.992	0.000
6	MAX	6.145	6.000	0.220	6.060	6.000	0.090	6.028	6.000	0.040	6.016	6.000	0.024	6.012	6.000	0.020
	MIN	6.070	5.925	0.070	6.030	5.970	0.030	6.010	5.988	0.010	6.004	5.992	0.004	6.000	5.992	0.000
8	MAX	8.170	8.000	0.260	8.076	8.000	0.112	8.035	8.000	0.050	8.020	8.000	0.029	8.015	8.000	0.024
	MIN	8.080	7.910	0.080	8.040	7.964	0.040	8.013	7.985	0.013	8.005	7.991	0.005	8.000	7.991	0.000
10	MAX	10.170	10.000	0.260	10.076	10.000	0.112	10.035	10.000	0.050	10.020	10.000	0.029	10.015	10.000	0.024
	MIN	10.080	9.910	0.080	10.040	9.964	0.040	10.013	9.985	0.013	10.005	9.991	0.005	10.000	9.991	0.000
12	MAX	12.205	12.000	0.315	12.093	12.000	0.136	12.043	12.000	0.061	12.024	12.000	0.035	12.018	12.000	0.029
	MIN	12.095	11.890	0.095	12.050	11.957	0.050	12.016	11.982	0.016	12.006	11.989	0.006	12.000	11.989	0.000
16	MAX	16.205	16.000	0.315	16.093	16.000	0.136	16.043	16.000	0.061	16.024	16.000	0.035	16.018	16.000	0.029
	MIN	16.095	15.890	0.095	16.050	15.957	0.050	16.016	15.982	0.016	16.006	15.989	0.006	16.000	15.989	0.000
20	MAX	20.240	20.000	0.370	20.117	20.000	0.169	20.053	20.000	0.074	20.028	20.000	0.041	20.021	20.000	0.034
	MIN	20.110	19.870	0.110	20.065	19.948	0.065	20.020	19.979	0.020	20.007	19.987	0.007	20.000	19.987	0.000

25	MAX	25.240	0.370	25.117	0.169	25.053	25.000	0.074	25.028	0.041	25.021	25.000	0.034
	MIN	25.110	0.110	25.065	0.065	25.020	24.979	0.020	25.007	0.007	25.000	24.987	0.000
30	MAX	30.240	0.370	30.117	0.169	30.053	30.000	0.074	30.028	0.041	30.021	30.000	0.034
	MIN	30.110	0.110	30.065	0.065	30.020	29.979	0.020	30.007	0.007	30.000	29.987	0.000
1	MAX	1.000	0.006	0.996	0.002	0.994	1.000	0.000	0.986	−0.008	0.982	1.000	−0.012
	MIN	0.994	−0.010	0.986	−0.014	0.984	0.994	−0.016	0.976	−0.024	0.972	0.994	−0.028
1.2	MAX	1.200	0.006	1.196	0.002	1.194	1.200	0.000	1.186	−0.008	1.182	1.200	−0.012
	MIN	1.194	−0.010	1.186	−0.014	1.184	1.194	−0.016	1.176	−0.024	1.172	1.194	−0.028
1.6	MAX	1.600	0.006	1.596	0.002	1.594	1.600	0.000	1.586	−0.008	1.582	1.600	−0.012
	MIN	1.594	−0.010	1.586	−0.014	1.584	1.594	−0.016	1.576	−0.024	1.572	1.594	−0.028
2	MAX	2.000	0.006	1.996	0.002	1.994	2.000	0.000	1.986	−0.008	1.982	2.000	−0.012
	MIN	1.994	−0.010	1.986	−0.014	1.984	1.994	−0.016	1.976	−0.024	1.972	1.994	−0.028
2.5	MAX	2.500	0.006	2.496	0.002	2.494	2.500	0.000	2.486	−0.008	2.482	2.500	−0.012
	MIN	2.494	−0.010	2.486	−0.014	2.484	2.494	−0.016	2.476	−0.024	2.472	2.494	−0.028
3	MAX	3.000	0.006	2.996	0.002	2.994	3.000	0.000	2.986	−0.008	2.982	3.000	−0.012
	MIN	2.994	−0.010	2.986	−0.014	2.984	2.994	−0.016	2.976	−0.024	2.972	2.994	−0.028
4	MAX	4.003	0.011	3.996	0.004	3.992	4.000	0.000	3.985	−0.007	3.981	4.000	−0.011
	MIN	3.991	−0.009	3.984	−0.016	3.980	3.992	−0.020	3.973	−0.027	3.969	3.992	−0.031
5	MAX	5.003	0.011	4.996	0.004	4.992	5.000	0.000	4.985	−0.007	4.981	5.000	−0.011
	MIN	4.991	−0.009	4.984	−0.016	4.980	4.992	−0.020	4.973	−0.027	4.969	4.992	−0.031
6	MAX	6.003	0.011	5.996	0.004	5.992	6.000	0.000	5.985	−0.007	5.981	6.000	−0.011
	MIN	5.991	−0.009	5.984	−0.016	5.980	5.992	−0.020	5.973	−0.027	5.969	5.992	−0.031
8	MAX	8.005	0.014	7.996	0.005	7.991	8.000	0.000	7.983	−0.008	7.978	8.000	−0.013
	MIN	7.990	−0.010	7.981	−0.019	7.976	7.991	−0.024	7.968	−0.032	7.963	7.991	−0.037
10	MAX	10.005	0.014	9.996	0.005	9.991	10.000	0.000	9.983	−0.008	9.978	10.000	−0.013
	MIN	9.990	−0.010	9.981	−0.019	9.976	9.991	−0.024	9.968	−0.032	9.963	9.991	−0.037
12	MAX	12.006	0.017	11.995	0.006	11.989	12.000	0.000	11.979	−0.010	11.974	12.000	−0.015
	MIN	11.988	−0.012	11.977	−0.023	11.971	11.989	−0.029	11.961	−0.039	11.956	11.989	−0.044
16	MAX	16.006	0.017	15.995	0.006	15.989	16.000	0.000	15.979	−0.010	15.974	16.000	−0.015
	MIN	15.988	−0.012	15.977	−0.023	15.971	15.989	−0.029	15.961	−0.039	15.956	15.989	−0.044
20	MAX	20.006	0.019	19.993	0.006	19.986	20.000	−0.001	19.973	−0.014	19.967	20.000	−0.020
	MIN	19.985	−0.015	19.972	−0.028	19.965	19.987	−0.035	19.952	−0.048	19.946	19.987	−0.054
25	MAX	25.006	0.019	24.993	0.006	24.986	25.000	−0.001	24.973	−0.014	24.960	25.000	−0.027
	MIN	24.985	−0.015	24.972	−0.028	24.965	24.987	−0.035	24.952	−0.048	24.939	24.987	−0.061
30	MAX	30.006	0.019	29.993	0.006	29.986	30.000	−0.001	29.973	−0.014	29.960	30.000	−0.027
	MIN	29.985	−0.015	29.972	−0.028	29.965	29.987	−0.035	29.952	−0.048	29.939	29.987	−0.061

Tolerance symbols

From Table 8-3 obtain   H9/d9

Basic size

From Table 8-2 select the preferred size of Ø 30

Hole size (H9)

From Table 8-4 obtain   Ø 30.000-30.052

Shaft size (d9)

From Table 8-4 obtain   Ø 29.883-29.935

Tightest fit

Smallest hole (MMC) − largest shaft (MMC)
30.000 − 29.935 = 0.065

Loosest fit

Largest hole (LMC) − smallest shaft (LMC)
30.052 − 29.883 = 0.169

Note that the 0.065 and 0.169 values are given in the table; therefore, it is not necessary to make the calculations. They are given as an aid to understanding the concept.

The **shaft basis system** uses the largest shaft (MMC) as the basic size; therefore, its upper deviation is 0. Although the shaft basis system is not as common, it does have applications in standard cold-finished shafting. A boring operation on the hole is needed to achieve its desired size.

A practical example will aid in understanding the system. The designer finds that a medium-drive fit will meet the criteria for two mating parts. Also, design calculations give a diameter of 15 mm. Tolerance symbols

From Table 8-3 obtain   S7/h6

Basic size

From Table 8-2 select the preferred size of 16

Hole size (S7)

   From Table 8-5 obtain   Ø 15.961-15.979

Shaft size (h6)

   From Table 8-5 obtain   Ø 15.989-16.00

Tightest fit

   Smallest hole (MMC) − largest shaft (MMC)

   $15.961 - 16.000 = -0.039$

Loosest fit

   Largest hole (LMC) − smallest shaft (LMC)

   $15.979 - 15.989 = -0.010$

In the shaft basis system, the fits are also given in the table.

   Before completing our discussion of mating parts, the concept of selective assembly should be explained. When precise parts (small tolerance) are needed for complicated devices to function properly, the demands on the processing equipment may be greater than their capability. When this occurs, there are two alternatives: (1) discard the parts that exceed the limits (expensive), or (2) use manual or computer-controlled selective assembly. In selective assembly, all parts are sorted and classified in various categories based on their size. Small shafts are then matched with small holes, medium shafts with medium holes, and so forth. In this manner, satisfactory fits that function properly can be obtained.

SURFACE TEXTURE

The proper functioning of mating parts as well as their service life are affected by the texture of the surface. **Surface texture** refers to repetitive or random deviations from the true surface. These deviations are described by roughness, waviness, lay, and flaws. They are illustrated in Figure 8-38.

   Roughness is the relatively finely spaced surface irregularities that are produced by the manufacturing process. **Roughness height** is the av-

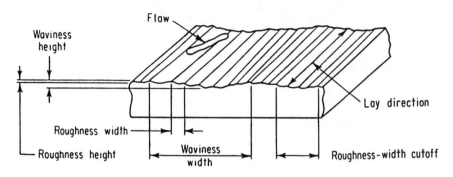

FIGURE 8-38
Characteristics of surface texture.

erage deviation from the mean plane of the surface. It is expressed in micrometers (microinches). **Roughness width** is the distance between successive peaks and is measured in millimeters (inches). **Roughness width cutoff** is the largest spacing of repetitive irregularities to be included in the measurement of the average roughness height. If it is not specified, a value of 0.8 mm (0.030 inch) is assumed.

Waviness is the widely spaced surface undulations that exceed the roughness irregularities as shown in Figure 8-38. Roughness may be viewed as superimposed on the wavy surface. **Waviness height** is the peak-to-valley distance between waves and is specified in millimeters (inches). **Waviness width** is the spacing between wave peaks or wave valleys and is measured in millimeters (inches). **Flaws** are irregularities such as cracks, checks, blowholes, scratches, and so forth, that occur at one place or at relatively infrequent or widely varying intervals on the surface. **Lay** is the direction of the surface pattern caused by the manufacturing process. Figure 8-39 shows six different lay patterns; also shown is the appropriate symbol.

A surface whose finish is to be specified is marked with the symbol as shown in Figure 8-40. The point of the "V" is on the line representing the surface, on the extension line, or on a leader that points to the surface. It is not necessary to use the entire symbol or specify all surface texture characteristics. The minimum basic symbol is shown in Figure 8-40(b). It is constructed with a 60° included angle and the right leg twice as long as the left. When the removal of material by machining is necessary, a horizontal bar is added at the top of the short leg as shown at (c). At (d), the added value in-

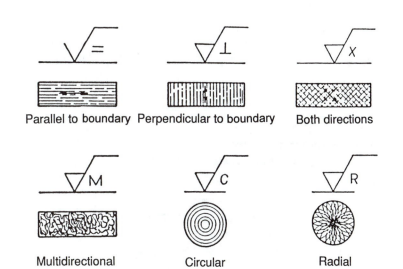

FIGURE 8-39
Lay patterns and symbols.

dicates the amount of stock in millimeters (inches) to be removed. When material removal is prohibited, a small circle is added in the "V" as shown at (e). This circle indicates that the surface is to be produced without removal of material by some production process such as die casting, powder metallurgy, injection molding, forging, or extruded shape.

Figure 8-41 shows the expected roughness height for some common manufacturing processes. Roughness height is sometimes specified in the symbol by two numbers—one on top of the other. In those cases, the values represent the maximum and minimum average roughness height.

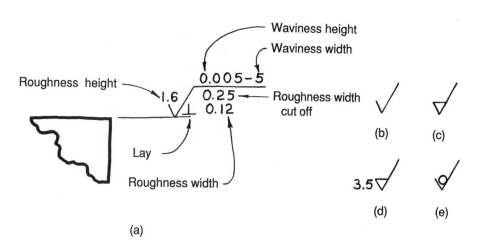

FIGURE 8-40
Surface texture symbols.

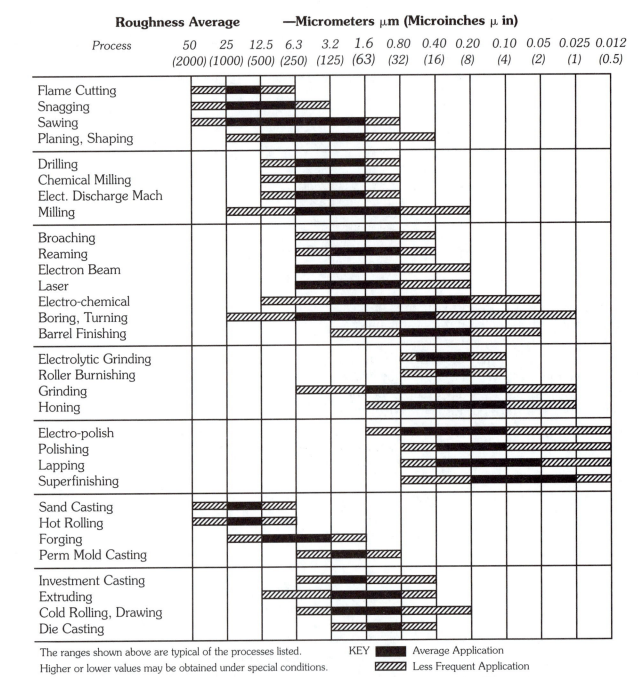

FIGURE 8-41
Roughness height of common processes.

SUMMARY

In summary, one must know or do the following:

1. Tolerance is the total amount a specific dimension is permitted to vary.

2. Limit dimensioning expresses the tolerance by giving an upper value and a lower value.

3. Plus and minus dimensioning expresses the tolerance as a deviation from a basic value. The tolerance is unilateral or bilateral.

4. Tolerance accumulation is minimized by the direct dimensioning technique and maximized by chain dimensioning.

5. Different manufacturing processes are capable of different degrees of accuracy.

6. Geometric tolerancing is described as a means of specifying the geometry of a part feature.

7. Specifying the geometric tolerance is simplified by the use of geometric characteristic symbols, datums, basic modifiers, and feature control frames.

8. A basic dimension is a theoretical exact dimension without tolerance.

9. There are three material condition modifiers: MMC, LMC, and RFS.

10. Rule 1 states that where only a feature of size is specified, the limits of size prescribe the amount of variation permitted in its form.

11. Rule 2 states that RFS applies with respect to the tolerance, the datum, or both, unless MMC is specified.

12. Datum planes are associated with the process or inspection equipment.

13. The datum reference frame is three mutually perpendicular planes.

14. The three planes are called the primary, secondary, and tertiary.

15. A primary datum plane requires three points of contact; a secondary datum plane requires two points of contact; and a tertiary datum plane requires one point of contact.

16. Datum targets are designated points, lines, or areas of contact. They are usually locating pins.

17. Flatness is the condition of a surface having all elements in one plane.

18. Straightness is the condition in which an element of a surface or an axis is a straight line.

19. Circularity is a condition of a surface of revolution where all points of the surface intersected by any plane perpendicular to a common axis or passing through a common center are equidistant from the center.

20. Cylindricity is a condition of a surface of revolution in which all points of the surface are equidistant from a common axis.

21. Profile tolerancing is a method of controlling irregular surfaces, lines, arcs, or unusual shapes as well as regular shapes.

22. Angularity is the condition of a surface or axis at a specified angle from a datum plane or axis.

23. Perpendicularity is the condition of a surface, center plane, or axis at a right angle to a datum plane or axis.

24. Parallelism is the condition of a surface equidistant at all points from a datum plane or an axis equidistant along its length to a datum axis.

25. Positional tolerance defines a zone in which the center, axis, or center plane is permitted to vary from the exact location, which is referred to as its true position.

26. Where a hole is at MMC, its axis must fall within a cylindrical tolerance zone whose axis is located at the true position.

27. Concentricity is the condition where the axis of all cross-sectional elements of a surface of revolution are common to the axis of a datum reference.

28. Symmetry is a condition in which a feature or features are symmetrically disposed about the center plane of a datum feature.

29. Circular runout provides control of circular elements of a surface.

30. Total runout provides combination control of all surface elements.

31. Whenever two parts are designed to function together, they are mating parts.

32. The specification of a mating part requires the basic size, fundamental deviation, and international tolerance grade.

33. There are three types of fits: clearance, interference, and transition.

34. The hole basis system is the most common system, because standard drill bits, reamers, and broaches are used, whereas shafts can be machined to any size.

35. Surface texture refers to repetitive or random deviations from the true surface.

36. Surface texture is described by roughness, waviness, lay, and flaws.

37. Symbols are as follows:

⟂	Datum feature
⊖	Datum target
✕	Target point
◎	Concentricity
○	Circularity
Ⓜ	MMC
Ⓛ	LMC
Ⓟ	Projected tolerance zone
Ⓕ	Free state
Ⓣ	Tangent plane
—	Straightness
//	Parallelism
▱	Flatness
⌀	Cylindricity
⊕	Position
⟿	All around
◠	Profile surface
⌒	Profile line
=	Symmetry
⟷	Between
⟂	Perpendicularity
∠	Angularity
↗	Runout circularity
⫮	Runout total

FASTENING TECHNIQUES

CHAPTER

9

Upon completion of this chapter, the student is expected to:

- Identify the types of thread representation, the different series of screw threads, three thread classes, and external and internal thread types found on a drawing.

- Be able to read the specification of a screw thread as it is shown on a drawing.

- Identify the basic weld symbols, four types of welds, the six fundamental welds, and the five types of weld joints.

- Identify at least three techniques of fastening that may be considered nontraditional.

INTRODUCTION

Individual piece parts and subassemblies that make up an end product are joined together by different fasteners. The type of fastener depends on the function of the device and whether the fastener is removable or permanent. Typical removable fasteners are bolts and nuts, screws, pins, and keys; typical permanent fasteners are rivets, nails, staples, drive pins, adhesives, and welding. The intent of this chapter is to illustrate how the different fastening techniques are specified on a drawing. Standards have been developed for some of the fasteners, while for others a local or general note is required.

THREADED FASTENERS

Screw threads are features vital to our industrial life. They are designed for many applications. Three basic applications of screw threads are: (1) holding parts together, (2) adjusting parts with reference to each other, and (3) transmitting power.

A thread is a uniform helical groove cut on or in a cylinder or cone. Thread cutting on a lathe is one of the most exacting lathe operations. A single-point cutting tool is used in lathe operations; however, a threading die for external thread cutting and a standard tap and die set for internal thread cutting can be used. Threads may also be ground or rolled, using a special machine. There are many different types and forms of threaded fasteners. A knowledge of thread terms, thread form, thread representation, thread series, thread classes, and specification of screw threads is needed. Threaded fasteners are manufactured in standard types and sizes. These standard parts may be purchased at a nominal cost; specially made sizes and forms are much more expensive.

Threads may be either internal or external. Figure 9-1 shows an example of an external and an internal thread as they might appear on an orthographic drawing. External threads are those on a round shaft, and internal threads are those inside a round hole. The screw thread is in the form of a spiral around the shaft or in the hole, and these spirals must match if the two threads are to fit together.

(a) External thread

(b) Internal thread

FIGURE 9-1
Screw threads.

Terminology

Following are some of the common terms and definitions for screw threads as shown in Figure 9-2.

External or male thread. A thread on the outside of a cylinder or cone.

Internal or female thread. A thread on the inside of a hollow cylinder or bore.

Pitch. The distance from a given point on one thread to a similar point on a thread next to it, measured parallel to the axis of the cylinder. The pitch is equal to 1 divided by the number of threads per millimeter or number of threads per inch.

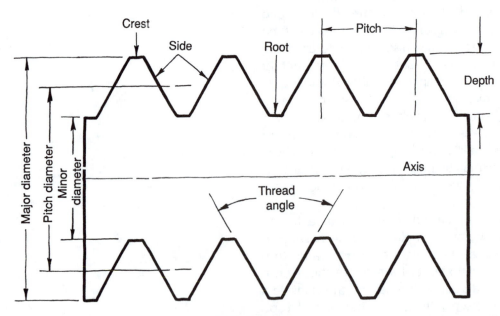

FIGURE 9-2
Nomenclature of screw threads.

Lead. The distance a screw thread advances axially in one complete revolution. On a single-thread screw, the lead is equal to the pitch. On a double-thread screw, the lead is equal to twice the pitch, and on a triple-thread screw, the lead is equal to three times the pitch (see Figure 9-6).

Crest (also called "flat"). The top or outer surface of the thread joining the two sides.

Root. The bottom or inner surface joining the sides of two adjacent threads.

Side. The side of a thread is the surface that connects the crest and the root.

Thread angle. The angle between the sides of adjacent threads, measured in an axial plane.

Depth. The depth of a thread is the distance between the crest and root of a thread measured perpendicular to the axis.

Major diameter. The largest diameter of a screw thread.

Minor diameter. The smallest diameter of a screw thread.

Pitch diameter. The diameter of an imaginary cylinder formed where the width of the groove is equal to one-half of the pitch. This is the critical diminsion of threading as the fit of the thread is determined by the pitch diameter.

Forms

The type of thread form specified by the design engineer is dependent on its application. Thread forms used in U.S. industry are based on approved U.S. standards. Figure 9-3 shows the profile of several commonly used threads. The unified form has been the standard in the English system; however, it is gradually being replaced by the metric form, which has slightly less thread depth. Sharp V threads are used to some extent where adjustment and holding power are essential. The knuckle thread is commonly found on light bulbs and bottle caps and is usually rolled from sheet metal. Acme and B&S worm threads are used for transmitting power.

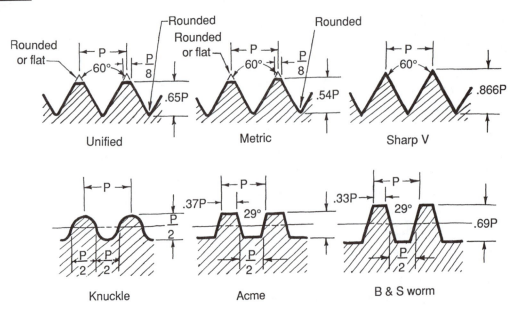

FIGURE 9-3
Thread forms.

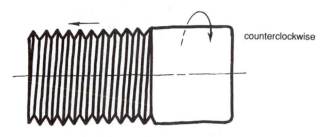

FIGURE 9-4
Right- and left-hand threads.

Left and Right Hand

A thread may be either right hand or left hand, depending on the application. Since most threads in use today are right hand, there is no indication to this effect on the drawing. If a thread is left hand, the letters LH are placed at the end of the thread note. Figure 9-4 shows a right- and left-hand thread as they would appear in detail on a drawing. Figure 9-5 shows the application of right-hand and left-hand threads. The thread must be such that tool pressure exerted on the grinding wheel will tend to tighten the nut holding the wheel on the shaft. If the thread is the wrong hand, the nut will loosen and the wheel will fall off. The same theory applies to the rear-axle nuts on an automobile, where one is right-handed and the other is left-handed. It also applies to bicycle pedals.

Lead

Lead is the distance a screw thread advances in one turn, measured parallel to the axis. Figure 9-6 shows the relationship of pitch to lead for single and multiple threads. In a single thread, the lead is equal to the pitch; in a double thread, the lead is equal to twice the pitch; and in a triple thread, the lead is equal to three times the pitch.

Figure 9-7 shows an application of multiple threads. In single threads the lead is equal to the

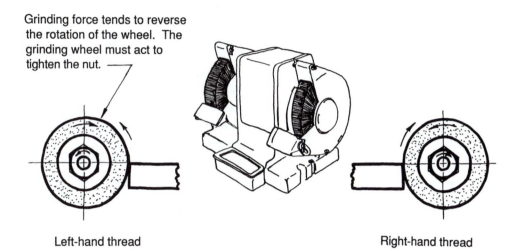

Grinding force tends to reverse the rotation of the wheel. The grinding wheel must act to tighten the nut.

Left-hand thread

Right-hand thread

FIGURE 9-5
Application of right- and
left-hand threads.

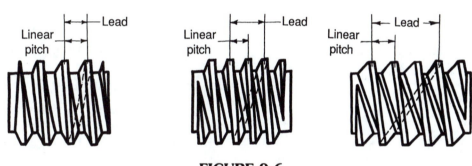

Lead

Linear pitch

Lead

Linear pitch

Lead

Linear pitch

FIGURE 9-6
Multiple screw threads.

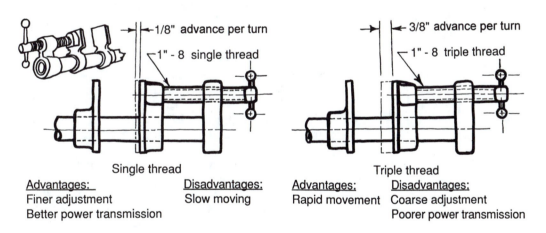

1/8" advance per turn

1" - 8 single thread

3/8" advance per turn

1" - 8 triple thread

Single thread

Triple thread

Advantages:
Finer adjustment
Better power transmission

Disadvantages:
Slow moving

Advantages:
Rapid movement

Disadvantages:
Coarse adjustment
Poorer power transmission

FIGURE 9-7
Application of multiple threads.

pitch, and in one turn of the screw, the screw advances one pitch. The pitch (P) is equal to 1 divided by the number of threads per inch. In the single-thread application, we have 8 threads per inch; therefore, with a complete turn of the handle, the clamping head will advance ⅛ inch. For the triple thread, a complete turn of the handle will advance the clamping head ⅜ inch. Single threads provide finer adjustment and better power transmission, while multiple threads provide rapid movement. Whenever double leads or triple leads are used, quick motion is obtained but not great power.

Representation

Threads may be represented on drawings using one of three conventional methods: detailed, schematic, and simplified. To save drafting time, the schematic or simplified conventions are used. The detailed convention is sometimes used to show the geometry of a thread form as a portion of a greatly enlarged detail on a drawing. Figure 9-8 illustrates the representation for external threads and for internal threads with a sectioned view.

Unified Series (English)

There are 11 standard series of threads listed under the unified national form. There are three series with graded pitches and eight series with constant

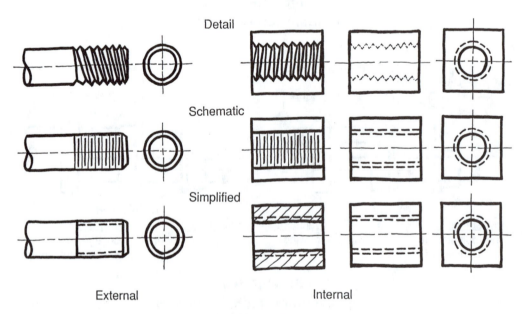

Detail

Schematic

Simplified

External

Internal

FIGURE 9-8
Screw thread representation.

pitches. In graded pitches the pitch gets smaller as the diameter increases whereas in constant pitches the pitch remains the same regardless of the diameter. Table 9-1 shows the series for basic major diameters between 0.0600 and 1.5000 inches.

The unified national coarse (UNC) thread series is a general-purpose thread for screws, bolts, and nuts where a finer thread is not required. The unified national fine (UNF) thread series has a greater number of threads per inch and is used where the length of the threaded engagement is short and a small lead angle is desired. This type of thread finds application in the aircraft and automotive industries. The unified national extra-fine series (UNEF) has application where even shorter lengths of thread engagement are required, such as for thin-walled tubes, nuts, ferrules, and couplings.

Of the eight series with constant pitch, 8 UN, 12 UN, and 16 UN are the most common. The general application of the 8-pitch series (8 threads per inch) is for bolts for high-pressure pipe flanges, cylinder head studs, and similar fasteners. The 12-pitch thread series (12 threads per inch) is used in boiler work and for thin nuts on shafts and sleeves in machine construction. The 16-pitch thread series (16 threads per inch) is used where a fine thread is necessary, regardless of the diameter.

Threads are further classified by the tolerance and allowance permitted in their manufacture. **Allowance** is the difference between the largest shaft (maximum material condition) and smallest hole (minimum material condition). It is the tightest fit. A class 1 fit is recommended for screw threads where a loose fit between mating parts permits frequent and quick assembly and disassembly of parts. A class 2 fit represents a high-quality type of thread product and is recommended for nuts, bolts, and screws used in normal applications found in mass-production industries. A class 3 fit has a tight fit allowance between mating parts and has closer tolerances than either a class 1 or 2 fit. This type of fit, because of its precision and cost, should only be used in special cases. The number for the class of fit is usually followed by the letter A or B, indicating an external (A) or internal (B) thread callout.

Metric Thread Series

There are 14 standard series of threads, and these are shown in Table 9-2 for diameters between 1.6

TABLE 9-1
Unified Standard Screw Thread Series

THREADS PER INCH

SIZES		BASIC MAJOR DIAMETER	SERIES WITH GRADED PITCHES			SERIES WITH CONSTANT PITCHES								SIZE
PRIMARY	SECONDARY		COARSE UNC	FINE UNF	EXTRA FINE UNEF	4UN	6UN	8UN	12UN	16UN	20UN	28UN	32UN	
0		0.0600	—	80	—	—	—	—	—	—	—	—	—	0
	1	0.0730	64	72	—	—	—	—	—	—	—	—	—	1
2		0.0860	56	64	—	—	—	—	—	—	—	—	—	2
	3	0.0990	48	56	—	—	—	—	—	—	—	—	—	3
4		0.1120	40	48	—	—	—	—	—	—	—	—	—	4
5		0.1250	40	44	—	—	—	—	—	—	—	—	—	5
6		0.1380	32	40	—	—	—	—	—	—	—	—	UNC	6
8		0.1640	32	36	—	—	—	—	—	—	—	—	UNC	8
10		0.1900	24	32	—	—	—	—	—	—	—	—	UNF	10
	12	0.2160	24	28	32	—	—	—	—	—	—	UNF	UNEF	12
1/4		0.2500	20	28	32	—	—	—	—	—	UNC	UNEF	UNEF	1/4
5/16		0.3125	18	24	32	—	—	—	—	—	20	28	UNEF	5/16
3/8		0.3750	16	24	32	—	—	—	—	UNC	20	28	UNEF	3/8
7/16		0.4375	14	20	28	—	—	—	—	16	UNF	UNEF	32	7/16
1/2		0.5000	13	20	28	—	—	—	—	16	UNF	UNEF	32	1/2
9/16		0.5625	12	18	24	—	—	—	UNC	16	20	28	32	9/16
5/8		0.6250	11	18	24	—	—	—	12	16	20	28	32	5/8
	11/16	0.6875	—	—	24	—	—	—	12	16	20	28	32	11/16
3/4		0.7500	10	16	20	—	—	—	12	UNF	UNEF	28	32	3/4
	13/16	0.8125	—	—	20	—	—	—	12	16	UNEF	28	32	13/16
7/8		0.8750	9	14	20	—	—	—	12	16	UNEF	28	32	7/8
	15/16	0.9375	—	—	20	—	—	—	12	16	UNEF	28	32	15/16
1		1.0000	8	12	20	—	—	UNC	UNF	16	UNEF	28	32	1
	1 1/16	1.0625	—	—	18	—	—	8	12	16	20	28	—	1 1/16
1 1/8		1.1250	7	12	18	—	—	8	UNF	16	20	28	—	1 1/8
	1 3/16	1.1875	—	—	18	—	—	8	12	16	20	28	—	1 3/16
1 1/4		1.2500	7	12	18	—	—	8	UNF	16	20	28	—	1 1/4
	1 5/16	1.3125	—	—	18	—	—	8	12	16	20	28	—	1 5/16
1 3/8		1.3750	6	12	18	—	UNC	8	UNF	16	20	28	—	1 3/8
	1 7/16	1.4375	—	—	18	—	6	8	12	16	20	28	—	1 7/16
1 1/2		1.5000	6	12	18	—	UNC	8	UNF	16	20	28	—	1 1/2

TABLE 9-2
Metric Screw Thread Series

NOMINAL SIZE (MM)	GRADED PITCHES		THREE-DIAMETER PREFERENCE			CONSTANT PITCHES (MM)											
	COARSE	FINE	1	2	3	6	4	3	2	1.5	1.25	1	0.75	0.5	0.35	0.25	0.2
1.6	0.35	—	1.6	—	—	—	—	—	—	—	—	—	—	—	—	—	0.2
1.8	0.35	—	—	1.8	—	—	—	—	—	—	—	—	—	—	—	—	0.2
2	0.4	—	2	—	—	—	—	—	—	—	—	—	—	—	—	0.25	—
2.2	0.45	—	—	2.2	—	—	—	—	—	—	—	—	—	—	—	0.25	—
2.5	0.45	—	2.5	—	—	—	—	—	—	—	—	—	—	—	0.35	—	—
3	0.5	—	3	—	—	—	—	—	—	—	—	—	—	—	0.35	—	—
3.5	0.6	—	—	3.5	—	—	—	—	—	—	—	—	—	—	0.35	—	—
4	0.7	—	4	—	—	—	—	—	—	—	—	—	—	0.5	—	—	—
4.5	0.75	—	—	4.5	—	—	—	—	—	—	—	—	—	0.5	—	—	—
5	0.8	—	5	—	—	—	—	—	—	—	—	—	—	0.5	—	—	—
5.5	—	—	—	—	5.5	—	—	—	—	—	—	—	—	0.5	—	—	—
6	1	—	6	—	—	—	—	—	—	—	—	—	0.75	—	—	—	—
7	1	1	—	—	7	—	—	—	—	—	—	—	0.75	—	—	—	—
8	1.25	1	8	—	—	—	—	—	—	—	—	1	0.75	—	—	—	—
9	1.25	—	—	—	9	—	—	—	—	—	—	1	0.75	—	—	—	—
10	1.5	1.25	10	—	—	—	—	—	—	—	1.25	1	0.75	—	—	—	—
11	1.5	—	—	—	11	—	—	—	—	—	—	1	0.75	—	—	—	—
12	1.75	1.25	12	—	—	—	—	—	—	1.5	1.25	1	—	—	—	—	—
14	2	1.5	14	—	—	—	—	—	—	1.5	1.25	1	—	—	—	—	—
15	—	—	—	—	15	—	—	—	—	1.5	—	1	—	—	—	—	—
16	2	1.5	16	—	—	—	—	—	—	1.5	—	1	—	—	—	—	—
17	—	—	—	—	17	—	—	—	—	1.5	—	1	—	—	—	—	—
18	2.5	1.5	—	18	—	—	—	—	2	1.5	—	1	—	—	—	—	—
20	2.5	1.5	20	—	—	—	—	—	2	1.5	—	1	—	—	—	—	—
22	2.5	1.5	—	22	—	—	—	—	2	1.5	—	1	—	—	—	—	—
24	3	2	24	—	—	—	—	—	2	1.5	—	1	—	—	—	—	—
25	—	—	—	—	25	—	—	—	2	1.5	—	1	—	—	—	—	—
26	—	—	—	—	26	—	—	—	—	1.5	—	1	—	—	—	—	—

and 26 mm. Note that thread diameters are selected based on an order of preference using columns 1, 2, and 3 in that order.

Threads are further classified by tolerance grades for the crest (major diameter for external and minor diameter for internal) and pitch diameter using the numbers 3 through 9, with 6 being nearly equal to the 2A and 2B classes of fit for the unified system. Grades with numbers less than 6 are used for fine fits and short lengths of engagement; grades with numbers greater than 6 are used for coarse fits and long engagements. Letters are used to specify the allowance and position (internal/external). For internal threads, G = small allowance and H = no allowance; for external threads, e = large allowance, g = small allowance, and h = no allowance.

Designation

Screw threads are made to specific dimensions to work properly. To interpret thread dimensions correctly, you must know how threads are specified. The unified, metric, acme, and pipe are the most common thread forms; therefore, this discussion will cover these four.

Unified. The specifications of a unified thread are shown with a series of numbers and letters. The pattern for a unified series is shown in Figure 9-9 and must be read from left to right. The first number (¼) shows the nominal size (major diameter) of the thread. This number represents the fractional diameter, the decimal diameter, or the number of the screw size. The second number (20) indicates the number of threads per inch. The thread series (UNC) is the next group of letters. In this example, unified coarse is indicated. The last number and letter (2A) indicates the class of fit, and whether the thread is internal or external. The letters used are A and B. The A means the thread is external and the B indicates an internal thread.

At other times other information is added to the standard designation as shown in Figure 9-9(b). The LH means that the thread is left hand. Threads are considered to be right hand unless an LH designation is used. The last entry (2.50) indicates the length of thread.

Metric. A series of numbers and letters are also used to designate metric threads. Figure 9-10

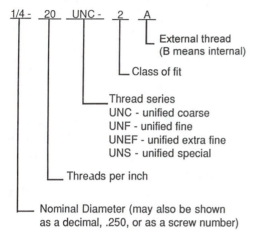

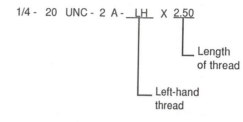

(a) Basic description

(b) Additional information

FIGURE 9-9
Unified thread designation.

shows a complete callout for a metric thread. The letter M for metric followed by the diameter in millimeters (10) and the pitch in millimeters (1.5) separated by the "3" sign. Next is the tolerance grade (6) and the tolerance position (g), which states that it is an external thread with small allowance. Finally comes the length of engagement symbol: S for short, N for normal, and L for long. Where the crest and pitch diameter tolerance grades are different, both must be specified. In many cases only the diameter and pitch are specified.

Acme. An acme thread follows the same basic format as a unified thread. Acme threads are designated by their nominal diameter, threads per inch, and class fit (Figure 9-11). The principal differences are the thread classes and the word acme. The thread classes 2, 3, 4, and 5 are commonly used for acme threads, with a class 2 the loosest and a class 5 the tightest. A general-purpose acme thread is indicated by the uppercase letter G and centralizing acme threads by the letter C. A stub acme thread is designated, for example, ⅜-12 stub acme without a letter symbol.

Pipe. There are three forms of pipe threads used in industry: regular, dryseal, and aeronautical. The regular pipe thread is the standard for the plumbing trade. Dryseal pipe thread is used for automotive, refrigeration, and hydraulic tube and pipe fit-

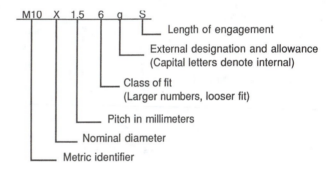

FIGURE 9-10
Metric thread designation.

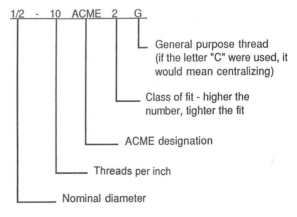

FIGURE 9-11
Acme thread designation.

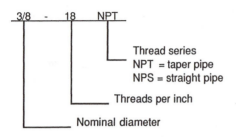

FIGURE 9-12
Pipe thread designation.

tings. The standard in the aerospace industry is aeronautical pipe thread. The American standard pipe threads specifications are listed in sequence on a print, by nominal size, number of threads per inch, and the symbol for the thread series and form (Figure 9-12). NPT (American standard taper pipe thread) specifies a taper pipe thread and NPS specifies a straight pipe thread.

Typical thread notes for an internal and external feature using English units are shown in Figure 9-13 and metric units in Figure 9-14. In the internal simplified metric, the middle representation illustrates a situation where the thread goes to the bottom of the hole. In the right representation the

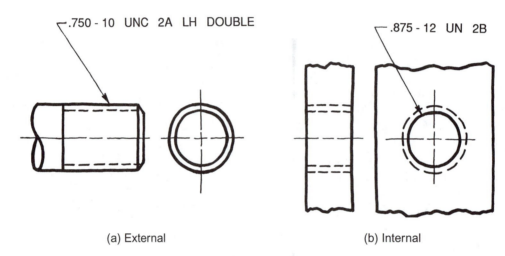

(a) External　　　　　　　　　　(b) Internal

FIGURE 9-13
Unified thread specifications.

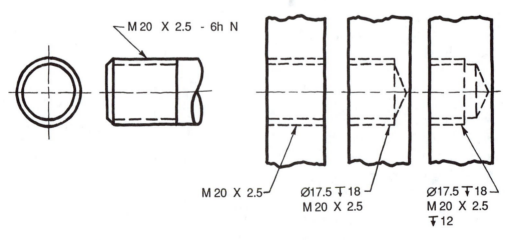

(a) External　　　　　　　　　　(b) Internal

FIGURE 9-14
Metric thread specifications.

hole is 18 mm deep and the threads 12 mm deep, which is easier to manufacture.

Relief

Some drawings may show a note calling for thread relief. The purpose of thread relief is to relieve the material so that the threading die may pass over the relief and make the threads complete to the end of the piece, thereby allowing the nut to be screwed tightly against the shoulder. One method to state thread relief is by the width and the depth of groove, as shown in Figure 9-15.

Standard Fasteners

Threads are used to make standard fasteners such as bolts, nuts, cap screws, machine screws, and setscrews. In the specification, the thread information is given first, followed by the fastener information.

The most common fastener is the bolt and nut, and they are shown in Figure 9-16. Heads may be regular or heavy depending on application

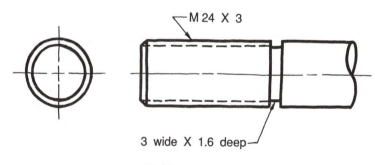

**FIGURE 9-15
Thread relief.**

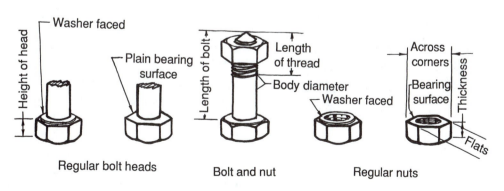

**FIGURE 9-16
Bolt and nut.**

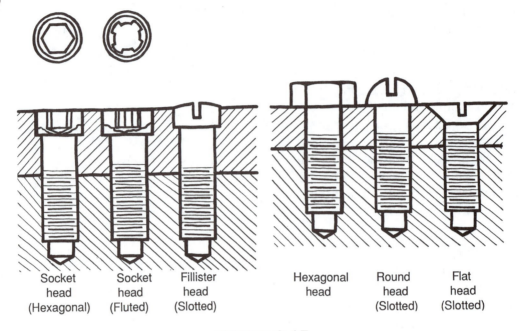

Socket head (Hexagonal) Socket head (Fluted) Fillister head (Slotted) Hexagonal head Round head (Slotted) Flat head (Slotted)

FIGURE 9-17
Cap screws.

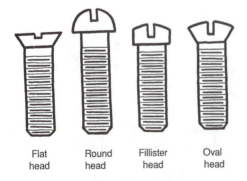

Flat head Round head Fillister head Oval head

FIGURE 9-18
Machine screws.

and are usually hexagonal rather than square in shape. Finished bolt heads and nuts are machined on the bearing surface to provide a washer face for the bolt head or nut.

Another common fastener is the cap screw and various types are shown in Figure 9-17. They are used to join two parts together. One of the parts is an oversized hole and the other is threaded and acts as a nut. When the cap screw is tightened, the two parts are securely held together. Different types of cap screws provide for above- or below-surface mounting. Flat head screws are useful when a flat surface is required above the part.

Machine screws are similar to cap screws and are shown in Figure 9-18. They are particularly adapted to screwing into thin materials, and the smaller screws are threaded to the head. Machine screw nuts are used mainly with the roundhead and flathead types and are hexagonal in shape.

Setscrews are used principally to prevent motion between two parts, such as a pulley and a shaft. The setscrew is screwed through one part, and the point presses on the other. As shown in Figure 9-19, there are six different points and three different head styles.

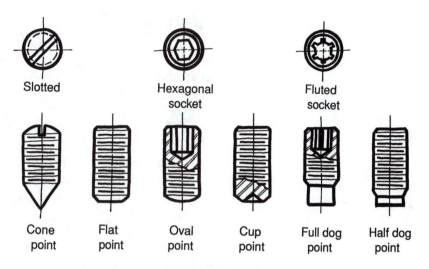

**FIGURE 9-19
Setscrews.**

RIVETS

One example of a permanent fastener is the rivet. Figure 9-20 illustrates the use of permanent fasteners in the form of rivets. The size of the rivet is designated by body diameter and usage varies from small toys to bridges. Many different head styles are available to suit the application. In structural application, heading is usually done with a die held in a powered hammer and the rivet held up with an anvil or backup bar. Other applications may use a rivet machine, in which the two pieces to be joined have prepierced holes and are located over a mandrel. A rivet is then fed into position down a track and picks up the mandrel; the hammer drives the rivet through the two pieces, striking an anvil on the underside and forming a head, thus securing the two pieces together.

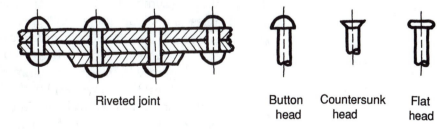

**FIGURE 9-20
Rivets.**

WELDING

Another example of a permanent type of fastener is welding. Welding processes are classified into three main types: (1) arc welding, (2) gas welding, and (3) resistance welding. These main types can be further broken down into subtypes. For example, in arc welding we find such types as metal arc, TIG (tungsten electrode used with an inert gas), MIG (metal inert gas welding), and others. The most common type of gas welding is oxygen and acetylene, which uses a high-temperature flame. Under the resistance welding classification are spot, seam, projection, and high-frequency welding, to name a few.

Two other types of metal joining that are closely allied to welding are brazing and soldering. Brazing is the process of joining metals with a non-ferrous filler metal that has a melting point below that of the metals being joined. The American Welding Society (AWS) defines the melting point of the filler as above 427°C (800°F). Below this temperature are the solders. In soldering, two or more pieces of metal are joined by means of a fusible alloy or metal, called solder, which is applied in the molten state.

The AWS has developed and adopted standard procedures for using symbols to indicate location, size, strength, geometry, and other information necessary to describe the weld required. In Figure 9-21 is the welding symbol with the standard placement of information concerning the weld. A full understanding of this standard symbol is not expected, since only a few will have an occasion to use it to make sketches or to interpret the sketch or drawing with the welding symbols. It is expected that, upon seeing the basic welding symbol, you will recognize it. Each item of the basic welding symbol has its place and purpose and identifies some operation or part of the welding process. The symbol in its entirety is seldom necessary. It is modified for particular applications.

Figure 9-22 illustrates common welds, such as the fillet weld, lap weld, and butt weld. The specification of the type of weld and the amount of weld are very important in controlling the weld joint.

The welding processes that we have discussed thus far are some of the tried-and-true methods that have been used for years. However, technological

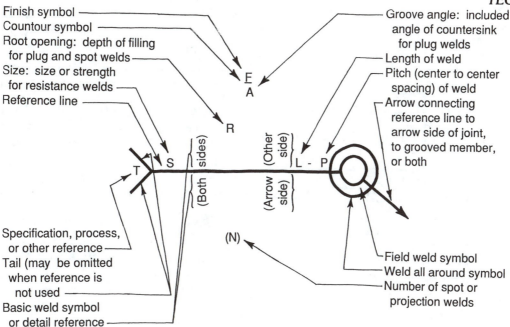

FIGURE 9-21
Basic weld symbols.

advances have produced many new welding processes. These may be classified into three areas: fusion, resistance, and solid state. Two specialized processes in the fusion area are electron-beam welding and laser welding. Welding processes that may be classified as solid-state processes are ultrasonic, explosive, diffusion, and friction processes.

OTHER TECHNIQUES

Several other techniques of fastening are as follows:

1. **Keys.** Basically, keys are a type of fastener that is used to prevent rotation of gears, pulleys, and collars on rotating shafts. The key is a piece of metal that fits into a keyseat in the shaft and keyway in the hub as shown in Figure 9-23.

2. **Locking devices.** Figure 9-24 shows several commonly used locking devices. The castellated nut is slotted to receive a cotter pin that passes through the shaft and slots in the nut. In the center of the figure a jam nut is used. The regular nut is tightened; then the

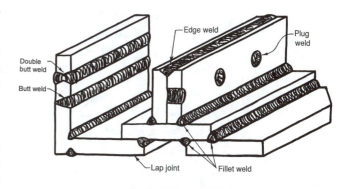

FIGURE 9-22
Types of welds.

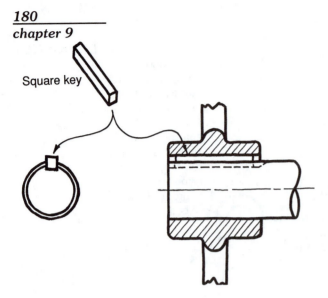

FIGURE 9-23
Square key.

jam nut is tightened against the regular nut. This creates a tension in the threads, tending to prevent the assembly from working loose. In the figure on the right a spring lock washer is shown. The reaction by the lock washer tends to prevent the nut from turning.

3. **Special types of bolts and screws.** Figure 9-25 shows some special types of bolts, screws, nuts, and fasteners that are used in fastening. One of the most common is the sheet metal or self-tapping screw, which is found in automobiles and appliances.

4. **Staples.** We are all familiar with the stapler used in offices, homes, and stores for the permanent joining of paper. This technique of joining materials is also used in manufacturing. Metal stitching uses the same technique as stapling paper: form the staple legs from a

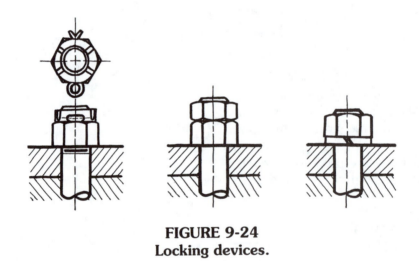

FIGURE 9-24
Locking devices.

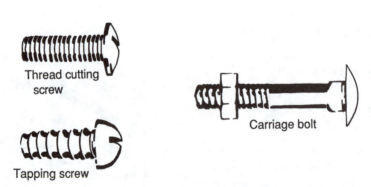

Thread cutting screw

Tapping screw

Carriage bolt

FIGURE 9-25
Special fasteners, bolts, and nuts.

wire roll, drive them through the material to be joined, and clench them against an anvil on the opposite side.

5. **Adhesives.** In its never-ending search to make airplanes lighter, the aircraft industry is using chemical fastening techniques. Structural bonding of parts is also used extensively in automobiles, appliances, and other products. For example, structural adhesives have been used for years to bond brake linings to brake shoes. The bond develops strengths of 10,000 pounds in shear compared to 2500 pounds developed by the rivets formerly used.

SUMMARY

In summary, one must know or do the following:

1. Threaded fasteners, rivets, welding, pins, and adhesive bonding are some of the techniques used to join materials together, either permanently or temporarily.

2. Threaded fasteners are specified by the type of thread, size of thread, pitch, and tolerance.

3. Unified, acme, and metric are the three most common types of thread forms.

4. Threads may be either right hand or left hand, external or internal.

5. Lead is the distance a screw thread advances axially in one complete revolution.

6. Common thread terms are pitch, crest, root, depth, angle, major diameter, minor diameter, and pitch diameter.

7. Threads may be represented on a sketch or drawing in one of three conventional ways: detailed, schematic, or simplified. The preferred method is simplified.

8. Threads are presented on a sketch using a symbol designation.

9. Standard fasteners are bolts, nuts, cap screws, machine screws, and setscrews.

10. The three main welding processes are arc, gas, and resistance. Brazing and soldering are closely allied to welding.

11. A welding symbol with the standard placement of information concerning the weld is found on subassembly or assembly drawings.

12. Types of welds are butt, edge, fillet, plug, lap, and double butt.

13. Other types of welding processes are electron beam, laser, ultrasonic, explosive, diffusion, and friction.

14. Other fastening techniques are keys, locking devices, sheet-metal screws, staples, and adhesives.

15. The fastening technique used is specified on a subassembly or assembly drawing by graphical representation, symbols, local notes, or a general note.

WORKING
DRAWINGS

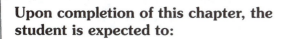

CHAPTER
10

Upon completion of this chapter, the
student is expected to:

- **Know the different types of drawing.**

- **Be able to identify the different parts
  of the title block.**

- **Be able to identify the different parts
  of the revision block and know its
  importance.**

- **Be able to identify the different parts
  of the bill of materials and know its use.**

- **Know the different types of notes and
  where each is used.**

- **Be able to interpret a working drawing.**

INTRODUCTION

Sketching or drawing is the universal language used by engineers, designers, technologists, technicians, and skilled craftsmen. Whether the drawing is made freehand or by a computer, it is used to convey all necessary information to personnel who will fabricate and assemble the building, aircraft, washing machine, camera, and so on. Copies of drawings are referred to as blueprints, prints, or simply drawings. The color of the copy is a function of the reproduction process.

Reading a drawing involves visualization and interpretation. Visualization is the ability to "see" or envision the size and shape of the object. You may wish to review Chapters 3 through 6, which discuss the different types of views. Operating personnel must be able to interpret symbols, dimensions, notes, and other information.

TYPES OF DRAWINGS

When a machine is manufactured or a structure built, the engineering department provides a set of drawings. These drawings are called working drawings and consist of two basic types: detailed drawings and assembly drawings.

A **detailed drawing** shows a single component or part. In a direct and simple manner, it gives the shape, size, material, finish, tolerance, and so on. The most common types are:

1. Machine drawings

2. Pattern drawings

3. Forging drawings

4. Sheet-metal drawings

5. Electronic drawings

6. Structural drawings

Detailed drawings are not to be confused with detailed views. A detailed view shows a part in the same plane and in the same arrangement, but in greater detail and on a larger scale than a principal view.

An **assembly drawing** shows the functional relationships of the various parts of a machine or structure as they fit together to make up an end product. The information provided on the assembly drawing is (1) the name of the assembly or subassembly, (2) the visual relationship of one part to another to ensure proper assembly of the various parts, and (3) the lists of parts showing number, name, and quantity. Assembly drawings are classified as follows:

1. Design assembly drawings

2. Subassembly drawings

3. Working assembly drawings

4. Checking assembly drawings

5. Installation drawings

6. Outline drawings

7. Illustrative assembly drawings

8. Exploded-view assembly drawings

9. Diagram assembly drawings

10. Plant layout drawings

COMPONENTS OF WORKING DRAWINGS

Working drawings are best analyzed by breaking them down into their components. The components are as follows:

1. Title block

2. Revisions

3. Bill of materials

4. General notes

5. Local notes

6. Size and shape description

Their usual locations on the drawing are illustrated in Figure 10-1.

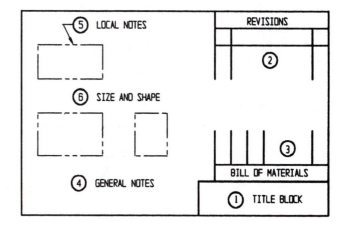

FIGURE 10-1
Working drawing components.

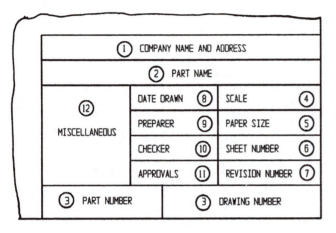

FIGURE 10-2
Title block.

TITLE BLOCK

The title block is illustrated in Figure 10-2; it is located in the lower right-hand corner of the drawing so that when folded properly the title block may be seen for easy reference and filing. Title blocks provide the basic information about the part or assembly. This information will vary from company to company depending on the specific needs of the company. It includes, but is not limited to, the following information:

1. Name and address of company

2. Name of part, subassembly, or assembly

3. Part number and drawing number, if different

4. Scale (When the equal sign is used, such as 1 = 4, the English system is implied; the colon, 1:4, implies the metric system.

5. Paper size or code letter (Table 10-1 gives the standard sizes and code letters for both systems.)

6. Sheet number (sheet 3 of 5)

7. Latest revision number (usually a code letter) and date

8. Date drawn

9. Preparer (designer or drafter)

10. Checker's initials and date

11. Approver's initials and date

12. Miscellaneous information, such as material, heat treatment, finish, estimated weight, and stress analysis

TABLE 10-1
Metric and U.S. Sheet Sizes

SHEET DESIGNATION	SHEET SIZE (MM)	CORRESPONDING U.S. SHEET SIZE	SHEET SIZE (INCHES)
A0	841 × 1189	E	34 × 44
A1	594 × 841	D	22 × 34
A2	420 × 594	C	17 × 22
A3	297 × 420	B	11 × 22
A4	210 × 297	A	8.5 × 11

REVISION BLOCK

The revision block is usually located in the upper right-hand corner of the drawing. It is shown in Figure 10-3 and is used to record all engineering changes to the original drawing. The revision block provides space for (1) the revision symbol, (2) description of the change, (3) date, and (4) approvals. For large, complicated drawings in which zoning is used to find a particular location, a zone column (5) is used.

Revisions or changes are made for a variety of reasons: improved design, change in manufacturing method, cost reduction, quality, and correction of errors in the original drawing. Whatever the reason, the revision must be documented. These revisions start at the top and proceed downward in a chronological order. A symbol, usually a letter, identifies the change in the revision block, and this same symbol is located adjacent to the change in the drawing.

It is very important that the record of changes be fully documented and the information disseminated to all appropriate personnel. This action will enable operating and inventory control personnel to perform their functions correctly. It is also important to enable service personnel to supply the correct part, knowing the serial number, model number, part number, or other identifying information.

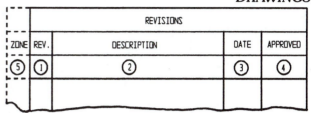

**FIGURE 10-3
Revision block.**

BILL OF MATERIALS

The bill of materials is shown in Figure 10-4. It usually appears immediately above the title block and contains a list of the parts and/or material used on or required by an assembly or installation drawing. The bill of materials is sometimes called parts list, materials list, list of materials, or schedule of parts. Sometimes the bill of materials is listed separately on a computer printout.

Accurate bills of material are extremely important for the production and inventory control system to function effectively. The bill of materials enables the purchasing department to order the quantity of materials needed to produce a given number of assemblies. It enables production control to requisition the required quantities from the storeroom and inventory control to maintain the correct quantity of items.

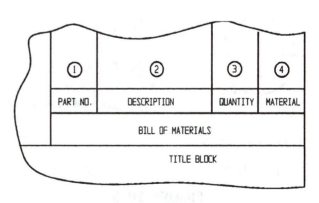

**FIGURE 10-4
Bill of materials.**

Although the information on the bill of materials varies from company to company, it usually contains the (1) part number, (2) description of material specification, (3) quantity required, and (4) the material. Additional information that may be added are industry code numbers and overall dimensions.

NOTES

Notes provide additional information and instructions in an efficient manner, thereby supplementing the graphical representation and the information found in the title block and revisions block. There are two types of notes: general notes and local notes, depending on the application. **General notes** apply to the entire drawing and are usually placed to the left of the title block in a horizontal position. Examples of general notes are:

REMOVE BURRS

ALL FILLETS AND RADII R4

ALL DIMENSIONS ±0.1 (mm)

Local notes or specific notes apply to certain features or areas. They are located near and directed to the feature or area by a leader, as shown in Figure 10-5a. Local notes can also be referenced from the field of the drawing using a leader and a note number enclosed in an equilateral triangle (sometimes called a **flag**). These are then listed in a convenient location on the drawing as shown in Fig. 10-5(b).

READING A DRAWING

A drawing cannot be read all at once any more than an entire page of print can be read at a glance. Both must be read a line at a time. Steps in reading a drawing are:

1. Read the title block.

2. Read the revisions.

3. Read the bill of materials (subassembly or assembly)

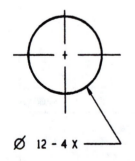

(a) Local note   (b) Flags

FIGURE 10-5
Examples of local notes.

4. Analyze the part or subassembly or assembly (whichever the case).

With this information in mind, let us begin to analyze a working drawing step by step.

TITLE BLOCK EXAMPLE

First we should check the part number and the drawing number, if different. As shown in Figure 10-6 the part number is 648167. The part or assembly name is descriptive, brief, and clearly stated. It starts with the basic name of the part or assembly, is followed by descriptive modifiers, and is read with the descriptive modifiers first. Therefore, the name for this part would be OUTER HARMONIC BALANCER WEIGHT DISK. The drawing scale indicates the size of the drawing as compared with

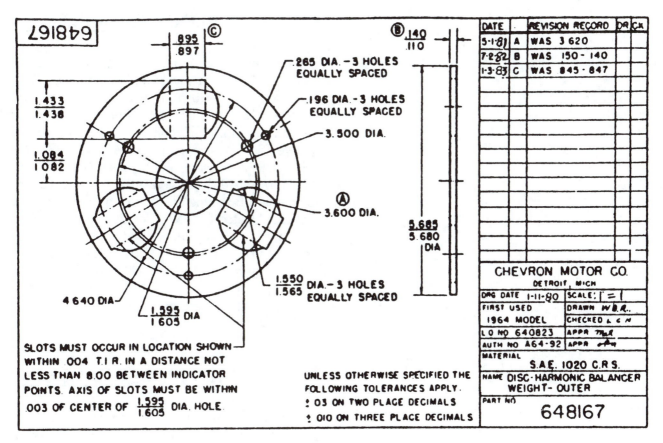

FIGURE 10-6
Title block and revision example.

the actual size of the part. The scale of 1 = 1 shows that the drawing is full size. The drawing data found in the title box are January 11, 1980, and that it was prepared by a person with the initials WBR.

Before a drawing is said to be finished it must be checked for completeness, accuracy, and clarity. Part number 648167 was checked by an individual with the initials LCN. You can see that two other individuals have given their approval. One of these was probably the drafting supervisor, the other perhaps the design engineer. Normally, before any drawing is used, these people initial and approve it. Before we fabricate a part, it is necessary that we know the material. To make this part, we find that the material callout is SAE (Society of Automotive Engineers) 1020 CRS (cold-rolled steel). Whether the steel will be sheet stock, strip stock, casting, and so forth, will be determined by the process engineer.

REVISION BLOCK EXAMPLE

Using Figure 10-6 again, we next analyze the revisions. Revision A involves changing the 3.620 dimension to 3.600. The description of this change, which is found under the revision record column, is short and concise. We can also see that revision A was made on May 1, 1981. On July 2, 1982, revision B was made, which affected the .150—.140 dimension of the part. If we look for revision B on the drawing of the part, we find that it is located at the top of the right side view. Notice the B has been circled and appears upper left from the .140—.110 dimension. What this tells us is that the stock thickness was changed from .150—.140 to .140—.110. Revision C changed the dimension from .845—.847 to .895—.897.

BILL OF MATERIALS EXAMPLE

A typical bill of materials is shown in Figure 10-7. First it is noted that one each of part numbers 1, 2, and 3 is required, that is, one base, one yoke, and one barrel. The material is cast iron; C.I. is the

Item	Qty.	Name	Material	Specifications
7	1	Cap Sc.	1112 ST.	M14 X 2 X 80 Hex
6	2	Hex. Nut	1112 ST.	M10 X 100 Jam
5	2	Set Sc.	1112 ST.	M6 X 1 X 12 SLT. Cone Pt. Set Sc.
4	1	Bushing	BRONZE	Oilite #1612
3	1	Barrel	C.I.	
2	1	Yoke	C.I.	
1	1	Base	C.I.	
Item	Qty.	Name	Material	Specifications

Bill of Materials

FIGURE 10-7
Bill of materials example.

abbreviation used. If our plant is a combination gray iron foundry and machining operation, we would be casting parts 1, 2, and 3 and then the necessary machine operations would follow. If, however, we do only machine operations, then the cast-iron parts would be purchased from a foundry, shipped to us, and then machined. Part 4 is a bushing made of bronze material with a specification of oilite no. 1612. You will note that only one is required and is purchased. Part number 5 is a setscrew made from 1112 steel with a diameter of 6 mm, 1 thread per millimeter, length of 12 mm, with a slotted head and a cone point. The quantity for part number 5 is 2 per unit assembly; therefore, if 100 unit assemblies were required, we would need 200 of part number 5. Part number 6 is a hexagonal jam nut with a diameter of 10 mm and a length of 100 mm, and part number 7 is a hexagonal head cap screw with a length of 80 mm and a diameter of 14 mm.

The bill of materials must contain sufficient information so that the parts may be manufactured or purchased complete without further information from the people who designed the device or prepared the drawing.

ANALYZE THE PART

In studying an orthographic projection drawing, certain rules of procedure will be helpful in visualizing an object:

1. Scan briefly all views.

2. Study carefully the front view for shape description.

3. Move from the front view to the other views and look for lines that describe the intersections of surfaces, the limits of a surface, or the edge view of a surface.

4. Study one feature at a time in the several views and begin to picture in your mind the shape of the real object.

One cannot read a blueprint by looking at a single view when more than one view is given. Two views will not always describe an object; and when three views are given, all three must be consulted to be sure the shape has been read correctly.

EXAMPLE PROBLEMS

Problem No. 1

Figure 10-8 is an illustration of a two-view detail drawing. The views shown are the front view and the top view. Our analysis should be as follows:

Step 1. Scan briefly all views shown.

Step 2. Study carefully the front view for shape description. The basic shape of this part is a rectangle that is 67 mm wide by 87 mm long. We do not know how thick the object is; therefore, we need to take the next step.

Step 3. Move from the front view to the top view. Here we see that the thickness or depth of this part is 18 mm.

Step 4. Study one feature at a time in the several views and begin to picture in your mind what the object will look like when all operations are complete.

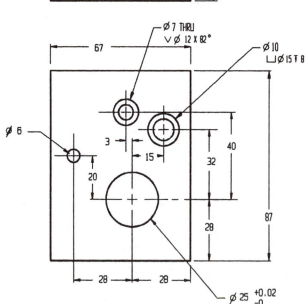

FIGURE 10-8
Visualizing the part.

As we look at the front view, we see that there are four holes of varying sizes. We have a 25-mm diameter hole with 0.02-mm tolerance. The center of this hole is located 28 mm from the right-hand edge of the part and 28 mm from the bottom edge. From all appearances, this hole goes clear through. We can check this by looking at the top view. If we project up from the outer edges of the hole in the front view to the top view, we will see hidden lines. These lines run from one edge (the front) to the other edge (the back) in the top view. Thus we see that the 25-mm-diameter hole goes clear through the part. It might be good to note here that the diameter of this hole is very precise. The machinist will drill this hole with a somewhat smaller drill size and then ream to give the final dimension of 25 to 25.02 mm. Note that the hole cannot be any smaller than 25 mm nor larger than 25.02 mm.

The location of the 6-mm hole can be found by looking at the front view and analyzing the dimensions given. We find that the hole is located 28 mm to the left and 20 mm above the large hole. This hole also appears to go clear through the part. We can verify this by projecting up from the front view to the top view. Again we find hidden lines that are as wide as the 6-mm hole, and they too go from front to back. Thus we see that this hole also goes through the part.

Now let us see how you can do on questions relating to the two remaining holes.

- What are the basic sizes of the two holes?

- What is the size of the counterbore? How deep is it?

- What is the degree callout for the countersink?

- What diameter will it be machined to?

- What is the distance of the 10-mm hole from the bottom edge?

- Is its vertical centerline located from the right-hand edge or the centerline of the 25-mm-diameter hole?

- What distance is the horizontal centerline of the 10-mm-diameter hole from the horizontal centerline of the 25-mm-diameter hole?

- Is the vertical centerline of the 7-mm-diameter hole 3 mm from the vertical centerline of the 25-mm-diameter hole?

- Which direction—right or left?

Problem No. 2

Figure 10-9 is a pattern drawing and a machining drawing. The pattern drawing will give information to the pattern maker as to size and configuration of the part prior to machining. The dimensions as shown on the pattern drawing are the casting size. The extra thick lines represent material that will be removed during the machining.

Our analysis of the pattern drawing should be as follows:

Step 1. Scan briefly all views shown. One view shows a rectangular-appearing object, while the other appears to be a series of concentric circles. Two circles are shown as hidden lines.

Step 2. Study carefully the front view for shape description. Upon analysis, we find a part that is 62 mm wide by 92 mm high.

Step 3. Move from the front view to the other view (top view). When we analyze the top view, we

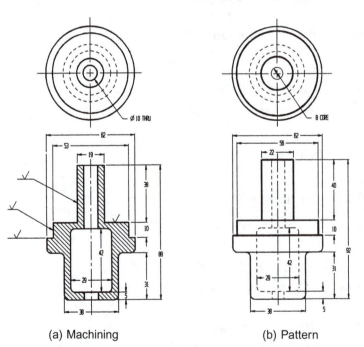

(a) Machining   (b) Pattern

FIGURE 10-9
Pattern and machining drawings.

find a series of concentric circles, the largest being 62 mm with a 8-mm core at the center.

Step 4.   Study one feature at a time in the views shown and begin to picture in your mind the shape of the real object. When we concern ourselves with the overall appearance of this part, we find that the series of concentric circles will be of varying diameters and lengths. For example, starting at the bottom we have a 38-mm-diameter circle that is 31 mm long. Above this we have a 62-mm diameter circle (this dimension is found between both views), which is 11 mm long (92—31—10—40). The 56-mm-diameter circle is 10 mm long and the 22-mm-diameter circle is 40 mm long.

When one analyzes a print, it is good practice to have a pencil available (preferably a red one) and to check off the dimensions as you analyze them.

Upon further analysis, we see hidden (dashed) lines within the part itself. We know from our previous units of instruction that hidden lines indicate edges, surfaces, and corners that are not visible in a particular view, such as the front view shown in the pattern drawing. The pattern maker, at this point in the analysis, would look at the machining drawing and verify the fact that the part contains a cavity. The machining drawing is a full-sectional view and shows the internal cavity that is to be made in the finished part. There will be times when a machining drawing will not be shown and the pattern drawing will show the sectional view. The internal cavity, as we will find, is of two different sizes. The larger cavity is 29 mm diameter by 42 mm long. At one end, it has a small hole of 8 mm diameter by 5 mm long; at the other end there is a 8-mm-diameter hole. The cavity found in this part will be made by placing a sand core in the mold. The molten metal will surround this core. Once the metal has solidified and the casting is removed from the mold and cleaned, we will find the cavity inside the part. This is the result of the core sand material breaking up and falling through the 8-mm-diameter openings.

Now let us return our attention to the shaded end (front view) of this pattern drawing. Perhaps your eye will focus on the 22 mm dimension. Earlier we had established the fact that this part was made with a series of concentric circles of varying lengths. Therefore, the 22 mm dimension is the diameter of the long projection. How long? Following the extension line at the top of the part out to the right, we find two arrowheads that end at this line. If we follow the arrowhead further from the part and follow it down vertically to its extreme end (at the bottom), we should have seen a figure of 92 mm. This tells us that the overall length of this part is 92 mm. But we want to know the length of the 22-mm-diameter projection. Let us go back up to the top extension line and move to the inner arrow. Following the dimension line down vertically to the next extension line and arrow combination, you should have passed a 40 mm dimension. Your interpretation should tell you that the 22-mm-diameter projection is 40 mm long.

What is the diameter and length of the next surface to be machined? As you look at the pattern drawing, you will note that the diameter is 56 mm by 10 mm long. The machining drawing gives the finished dimensions of the surfaces that are to be machined. These surfaces are identified by a √ on the drawing.

Problem No. 3

Figure 10-10 is a subassembly drawing of a hydraulic valve activator subassembly. As you know, most final assemblies consist of subassemblies and individual piece parts. As you look at this drawing you see that it is made up of three items, 2, 4, and 8. You will also see the relationship between the three as they are to be assembled. In this type of drawing, an outline and description of the parts are given so that the parts may be assembled in the proper order. A reference is also made as to the next assembly that it will be used on, assembly no. 14-4.

Figure 10-11 is known as a design layout drawing. This type of drawing will show the arrangement of parts, size, and the general overall relationship of the parts, one to another. This type of drawing is used to evaluate the overall design of the product. Once the design has been evaluated, it is then turned over to the detailer, who makes detailed drawings from the layout. As you can see in the drawing, the individual parts that make up the assembly are identified with a number that is not the drawing number. This number would be found on a bill of materials alongside the part number, part name, and so on. One part that is classed as a standard part, the flathead screw (no. 6), would not be detailed.

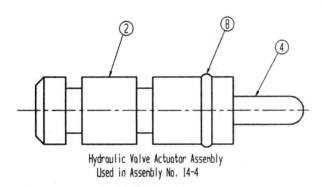

Hydraulic Valve Actuator Assembly
Used in Assembly No. 14-4

FIGURE 10-10
Subassembly drawing.

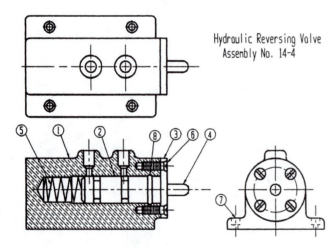

Hydraulic Reversing Valve
Assembly No. 14-4

FIGURE 10-11
Design layout drawing.

SUMMARY

In summary, one must know or do the following:

1. Working drawings consist of two basic types: assembly drawings and detail drawings.

2. Assembly drawings show the parts assembled in the unit or machine and their working relationships.

3. Assembly drawings serve as a place to show the assembled parts and to identify each part.

4. Assembly drawings seldom show all the details of the parts in the assembly.

5. A detail drawing illustrates a single component or part, showing a complete and exact description of its shape, dimensions, and features.

6. The parts of a working drawing are:
    a. Title block
    b. Revision block
    c. Bill of materials (for assembly drawings)
    d. Notes
    e. Detail part, subassembly, assembly

7. The procedure for reading a blueprint is:
    a. Read the title block.
    b. Read the revisions.
    c. Read the bill of materials.
    d. Analyze the part or subassembly or assembly.

CAD

CHAPTER
11

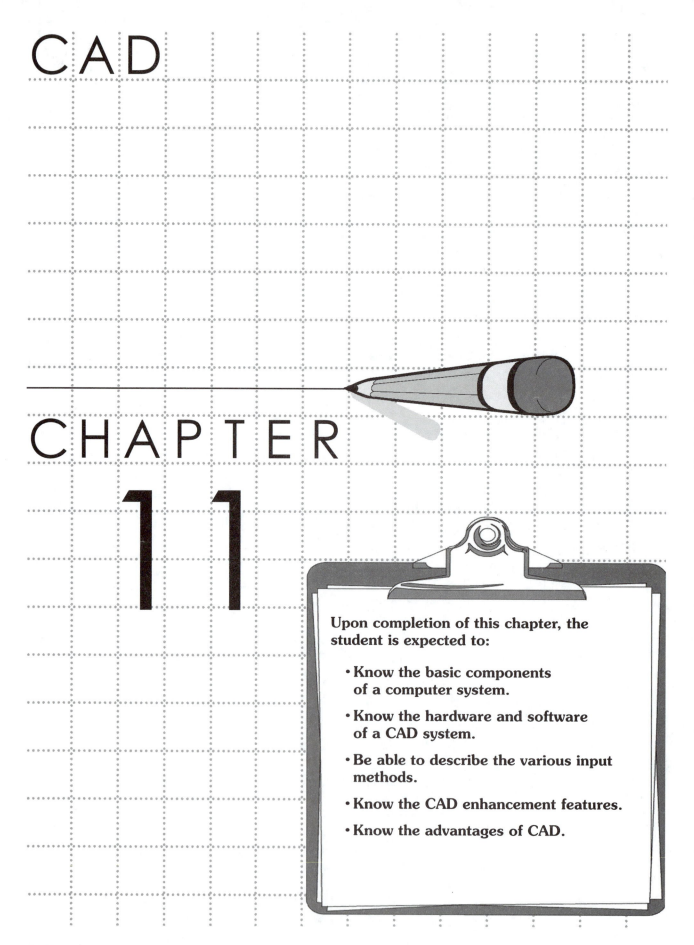

Upon completion of this chapter, the student is expected to:

- Know the basic components of a computer system.

- Know the hardware and software of a CAD system.

- Be able to describe the various input methods.

- Know the CAD enhancement features.

- Know the advantages of CAD.

INTRODUCTION

This book would not be complete without some mention of the acronym CAD/CAM, which stands for *computer-aided design and computer-aided manufacturing.* To put it simply, CAD/CAM is "the use of computers to aid the design and manufacturing process." The conceptualization of a new product, structure, or system, or the improvement thereof, must exist in the mind of the designer before it can become a reality. In most situations, the first step is to communicate ideas by means of a technical sketch.

In the development of an idea, many sketches as well as stress calculations, motion analysis, size of part, material specifications, production methods, and design layouts are made. It is the responsibility of the designer to oversee the preparation of part drawings and part specifications that provide for the many details of production, assembly, and maintenance of the product. To accomplish these tasks, the designer uses freehand sketches to record and communicate ideas rapidly to drafters and support personnel. The ability to convey ideas through sketches and the ability to work with a computer-controlled drawing system is essential.

The thrust of this book has been to familiarize the reader with the basics of technical drawing. Such terms as T-square, 30°-60° triangle compasses, templates, and so forth, were purposely avoided because the authors feel that like the slide rule, drawing instruments will be in limited demand in the years ahead. The ability to sketch, however, will always be needed. It is the age of computer technology and with it, CAD/CAM. The material presented in this chapter is, however, limited to information relating to CAD.

In the original development of CAD/CAM each was treated as a separate discipline and developed independently. Today, CAD/CAM integrates the design and manufacturing function rather than treating them as separate and distinct entities.

Engineering drawings have been and still are an integral part of industry. They are the link between design and manufacturing. Drawings, prepared according to prescribed drafting standards, quickly communicate to manufacturing personnel information for the eventual fabrication of a part or product. It is said, "A picture is worth a thousand words." Actually, a picture is worth much more since the speed of graphic communication can approach a rate of 50,000 times that of reading.

CAD DEFINED

Today it is not necessary to prepare an engineering drawing using conventional tools. Traditionally, drawing instruments have been used to apply India ink on vellum. Now the popular alternative is to prepare the drawing with the aid of a computer. This method is known as *computer-aided design* or *computer-aided design and drawing,* abbreviated CAD or CADD. It has rapidly replaced instrument drawings. Other terms used are *computer-assisted drawing, computer-augmented drawing,* and *computer-automated drawing.* These terms are used synonymously. They will be abbreviated as CAD throughout this chapter.

CAD HISTORY

The CAD/CAM process is an outgrowth of the development of **computer-numerical control** (CNC) machine tools. The start can be found in **numerical control** (NC), where an NC parts program was found to be a nucleus for a manufacturing data base. Users of computer-graphics system in the aircraft industry were generating shapes of complex geometries, with data being created in computer form. From this beginning, the obvious step was to utilize these data to generate manufacturing information. The coupling of these two activities became CAD/CAM—the marriage of computer-aided design and computer-aided manufacturing. Whenever computers are widely used for manufacture, the term CAM is applied to all procedures, where computers assist in the planning and production processes, in inventory control, and for the programming of NC machine tools.

CAD was introduced in 1964, when IBM made it commercially available. The first complete (turnkey) system was made available in 1970 by Applicon Incorporated. Only recently, however, has the dramatic impact of this new technology been felt. By the mid-1980s, CAD systems had become quite commonplace in manufacturing.

Toward the end of the 1990s, CAD systems are incorporated into the personal computer on every engineer's desktop.

THE CAD SYSTEM

Various combinations of equipment go to make up a CAD system. This holds true for small, medium-sized, and large system applications. The specific package selected depends largely on the needs of the user. Various types of drawings, referred to as **hard copy,** may be preferred by some, while others need no hard copy.

CAD systems are summarized as follows:

1. CAD systems are categorized as micro, mini, and mainframe.

2. Equipment for every CAD system will include processing, input, and output.

3. Processing includes programs stored on media and the means to drive them. The four types of media are: tape cartridges; floppy disks; hard disks; and CD-ROM, including optical read/write.

4. Common input equipment includes keyboard, digitizer/puck, and mouse. More recently, trackballs, pointing devices, and glidepoints have been introduced.

5. Output equipment is categorized as a display monitor (often referred to as a cathode ray tube), plotter, laser printer, inkjet printer, and CAM.

The state of the art in CAD is still evolving. Future systems will enable the user to "talk" to the computer. The ultimate goal of CAD includes a full implementation with CAM and a totally automated factory.

HARDWARE

CAD systems require hardware (i.e., a computer and peripheral devices such as terminals and plotters) and software (i.e., program of instruction that makes the hardware operate in the desired man-

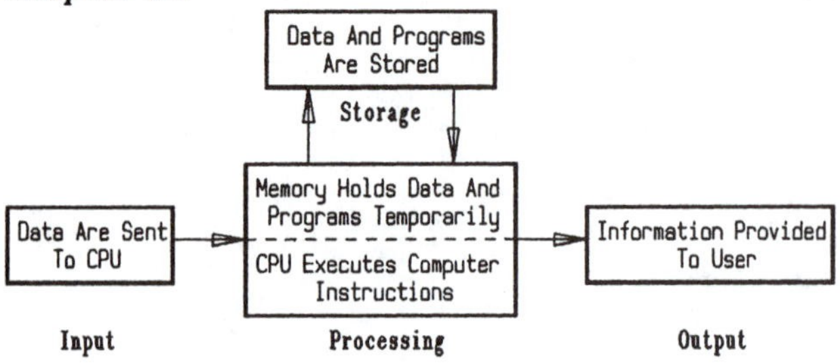

FIGURE 11-1
Computer system components.

ner). What follows is a brief discussion of some of the equipment that makes up a CAD system. It is not our intention to get into a long, drawn-out description of each piece of equipment but merely to familiarize you with some of the terms.

A **computer system** consists of three main areas of data handling—input, processing, and output—and is backed by a fourth, storage. These four components are shown in Figure 11-1.

Input devices take data in machine-readable form and sent it to the processing unit. The processor, more formally known as the central processing unit (CPU), manipulates input data into the desired information by executing computer instructions. It has a memory that temporarily holds data and instructions (programs). Output devices make the processed information available for use. Storage devices are auxiliary units outside the CPU that store additional data and programs.

CAD systems are categorized by type of computer. A **microcomputer** is a small, low-cost computer built around a microprocessor. These computers, which are often referred to as personal computers (PCs), have a single printed circuit board containing primary system chips. They are excellent for graphic applications where low price and small size are important. **Minicomputers** are smaller versions of large mainframe computers. Minicomputers have many of the capabilities of mainframe computers. The number of terminals that can be supported by a minicomputer varies from two to eight terminals. The large **mainframe** computer is distinguished by its capacity, function, and cost. Its main memory capacity is several orders of magnitude larger than the minicomputer and its computational speed is several times that of a mini- or microcomputer. In general, the larger the computer, the more sophisticated the CAD system and the higher the cost; however, a very adequate CAD system can be developed with a simple PC.

The **workstation** is the vehicle for communication with the CAD system. It represents a significant factor in determining how convenient and efficient it is for a designer to use the CAD system. The workstation must accomplish five functions:

1. Interface with the central processing unit (CPU)

2. Generate a steady graphic image for the user

3. Provide digital descriptions of the graphic image

4. Translate computer commands into operating functions

5. Facilitate communication between the user and the system

The use of interactive graphics has been found to be the best approach to accomplish these functions. A typical interactive graphics workstation consists of a display monitor, operator input devices, a computer, and sometimes output devices. A graphics design workstation showing some of these components is shown in Figure 11-2. The display monitor is the most crucial device in the system. By displaying letters and graphic representations of geometric shapes and design on the screen, the computer communicates with its human partner. In turn, he or she can "talk back" by using various input devices. This activity is possible because the display monitor and input devices are connected directly to the computer.

The alphanumeric **keyboard** is one of the primary input devices. The keyboard is mainly used for text and to increase the precision of dimensioning. It is the most common means of communicat-

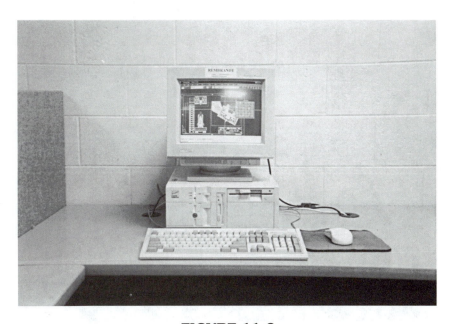

FIGURE 11-2
Workstation showing display monitor,
computer, keyboard, and mouse.

ing with the computer when operating outside the CAD mode. Figure 11-2 also shows a typical keyboard. When numerous workstations are located together in a room it is often called a **computer-aided design lab** or **CAD lab** as shown in Figure 11-3.

A **digitizer** is an electromechanical input device that resembles an electronic tablet and is shown in Figure 11-4. It allows direct input to a computer using data in the form of digital coordinates. The surface of the tablet corresponds to the face of the display monitor. A puck or light pen is used to transmit information to the computer. The digitizer or tablet is suitable for interaction since it allows the designer to work naturally with the puck or light pen and simulate freehand sketching. Often the tablet is divided into areas with significant meaning to the drafting system. These areas are called a **menu.** In effect, the menu gives access to a specific instruction or small drawing routine in the computer system. The use of the menu eliminates the need for typing coded instruction and simplifies the CAD system. Some versions of popular CAD software have the digitizer built into the display such that commands can be accessed easily via a mouse.

A command menu instructs the computer what to do, but it does not always specify where on

FIGURE 11-3
Computer-aided design lab.

FIGURE 11-4
Digitizer/puck with menu.

the display to execute the instruction. To specify this position, a **mouse** is used to move a cursor to any position on the face of the screen. Note that the puck, light pen, or keyboard can also be used to move the cursor. The mouse, shown in Figure 11-2, has a ball on its underside that rotates when the device is pushed along a surface. As the ball or wheels rotate, the cursor position on the graphics screen is determined. The mouse has buttons that provide for additional communication with the computer.

The **plotter** is the most common hard-copy output device for producing large-scale engineering drawings and is the peripheral that produces the finished drawing. There are several types of plotters. They are drum, flatbed, vertical bed, photo plotter, and electrostatic printer/plotter. One of the most common is the drum plotter, and it is shown in Figure 11-5. Drum plotters have paper partially wrapped around a drum that rotates, giving longitudinal paper movement under a pen carriage. The pen carriage traverses parallel to the axis of the drum, producing lateral pen movement.

A flatbed plotter is shown in Figure 11-6. They are used when increased accuracy is desir-

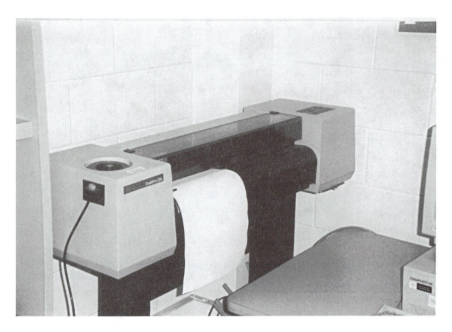

FIGURE 11-5
Drum plotter.

able. As the name implies, flatbed plotters operate by plotting on paper that rests on a flatbed. The pen moves in the X and Y directions across the bed. Vertical bed plotters, or easels, are an alternative to the flatbed type. This type of plotter is like a flatbed rotated through 90° so that it assumes the

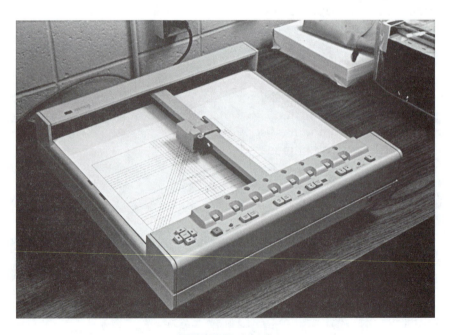

FIGURE 11-6
Flatbed plotter.

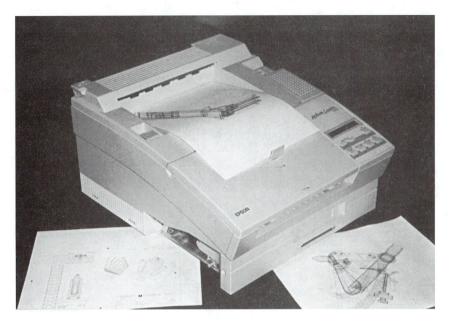

FIGURE 11-7
Laser printer.

position of a conventional drawing board. Flatbed plotters combine high speed with accuracy.

Another output device is the **printer,** and two are shown in Figures 11-7 and 11-8. Printers are much faster than plotters; however, some printers lack the accuracy of a plotter and cannot print large drawings. Dot-matrix printers are relatively inex-

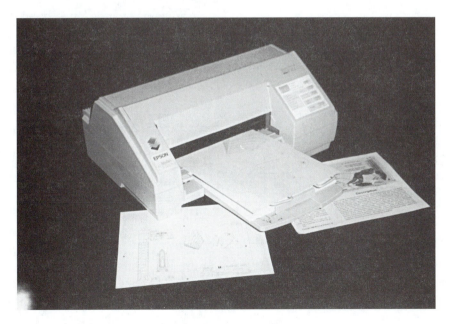

FIGURE 11-8
Color ink jet printer.

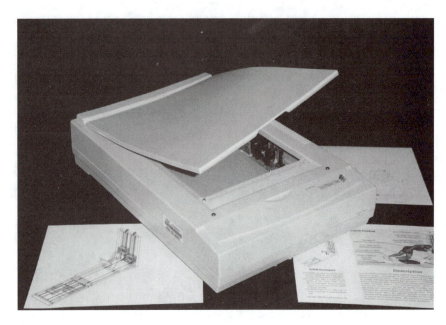

FIGURE 11-9
Color flatbed scanner.

pensive, but line quality is not as good as that of a plotter or laser printer. Laser printers have much better line quality; however, they are much more expensive than dot matrix printers. Recently, color ink jet printers have become a popular low-cost choice to obtain high-quality color outputs. Another very valuable peripheral device is a flatbed color scanner as shown in Figure 11-9. A scanner can be utilized to transfer drawings (even hand drawn) into a computer with the appropriate software. Text can also be scanned and interpreted with the aid of optical character recognition (OCR) software.

SOFTWARE

CAD hardware, or for that matter, any computer hardware would not be of much use unless there was suitable software to support it. Without such software the computer cannot function, "like a stereo system without any tapes." The term "software" is used to refer to all programs that run on a computer system to perform various tasks. In general, CAD software can be divided into three main categories: operating system, graphics system, and application system.

The **operating software** is the major system in a CAD system. It provides for such tasks as memory allocation, the scheduling of the processing unit, the operation of the input/output devices, and it arranges the priorities of operations. Most CAD systems have some form of auxiliary storage device, usually a disk drive, and one section of the operating system has to oversee the disk storage system. This part of the operating system is called the *disk operating system* (DOS) and it has the task of organizing the transfer of files between disk and memory as well as keeping a check on the location of files. The operating system software also acts as an interface between the user and the computer. This interface enables the user to communicate with the computer via the operating system language.

The function of the **graphics software** is to provide graphics capabilities. To meet the requirements of designers and drafters, graphics software must be able to accept, edit, revise, correct, and store data for primary graphic entities such as points, lines, planes, arcs, and circles as they combine to form multiview and pictorial representations.

Application software is software created for specific tasks to be performed in one of the technical fields, such as architectural, mechanical, structural, electronic, and highway design. It is the application software that turns the CAD hardware into a useful system. There are many different commercially available systems. CADKEY, AUTO-CAD, and VERSA-CAD are a few of the more common ones.

COMPUTER-AIDED DRAWING

Today a drawing may be made by using one or both of two basic techniques: instrument drawing or CAD. We have covered the basics of drawing, from the aspect of sketching. Points, lines, and surfaces have been used to describe objects by means of orthographic views, pictorials, sectional views, auxiliary views, tolerancing, and dimensioning. These fundamentals are essential for CAD.

It is at this point that one should be in a position to understand the unique coming together of a human being and a CAD system. The human being directs, provides the objectives, and makes the final evaluation. The CAD system performs the operations by supplying information and instructions as to how, when, and where operations are to be carried out.

In the modern CAD system the drawings are constructed interactively by the user supplying information to a computer via commands or with a menu-driven series of instructions. The drawings can easily be edited, manipulated, and stored for future retrieval. A plotter or printer is used to produce the final drawing to high standards of accuracy and line consistency.

INPUT METHODS

The keyboard is used primarily to input textual material, dimensions, and operating system commands. However, some of the applications software packages are designed to create the entire drawing with only keyboard entry. This method is not as efficient for geometric shapes as other input methods.

Some CAD systems use a large digitizer that resembles a traditional drawing board. Rough sketches are taped in place on the digitizer and traced electronically using a light pen or puck. Lines, arcs, circles, and other geometric shapes and forms are converted into electrical impulses and recorded as digital data in the computer's memory. The computer tidies up the drawing by straightening lines, smoothing out curves and arcs, and orienting lines at the proper angles. If desired, these data may be called up and displayed on the display monitor, where any necessary manipulations can take place.

In one form the digitizing tablet has an electronic grid system. Generators acting within the tablet produce discrete signals in response to the puck or light pen. A cross-hair symbol, called a cursor, that corresponds to the light-pen position will appear on the display monitor. It is by this means that x-y coordinates are created and stored in the computer's memory. Thus a drawing becomes a series of x-y coordinates.

Another input method is the mouse, which controls the cursor on the screen. A mouse has the same capability as a digitizer/puck, trackball, glidepoint, or light pen. However, it operates on any surface, with the relative position of the mouse ball determining the position of the cursor.

All of the many commands needed to communicate with most CAD systems are stored in memory and activated by a menu. Menus are displayed either on the screen or on the digitizer, as shown in Figure 11-4. The desired command can be selected by a variety of methods, such as keyboard, digitizer/puck or light pen, or mouse. Menus are hierarchically structured, as shown in Figure 11-10. To create a circle in this particular CAD system, you would first activate the CREATE function from the main menu. Next, the CIRCLE command is activated from the CREATE menu. Finally, the method of constructing the circle is activated by one of the CIRCLE commands, such as CTR + DIA, which stands for "center and diameter." The center may be located by the cursor or keyed in with the x-y coordinates. Of course, the diameter must be keyed in. A similar process is followed to dimension the circle. Within a short period of time, the user becomes quite proficient in using menus. In addition, prompts are given to make most systems quite user-friendly.

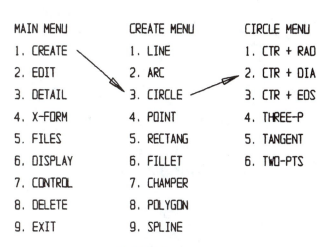

MAIN MENU	CREATE MENU	CIRCLE MENU
1. CREATE	1. LINE	1. CTR + RAD
2. EDIT	2. ARC	2. CTR + DIA
3. DETAIL	3. CIRCLE	3. CTR + EOS
4. X-FORM	4. POINT	4. THREE-P
5. FILES	5. RECTANG	5. TANGENT
6. DISPLAY	6. FILLET	6. TWO-PTS
7. CONTROL	7. CHAMPER	
8. DELETE	8. POLYGON	
9. EXIT	9. SPLINE	

FIGURE 11-10
**Hierarchically structured menu
for creating a circle.**

CAD ENHANCEMENT

CAD can do much more than merely duplicate those activities associated with instrument drawings. The abilities of a CAD system include some or all of the following functions: zoom, mirror, rotate, move, copy, crosshatch, automatic dimension, and spatial coordinates.

Zoom is the capability of the system to either enlarge or reduce a portion of a drawing displayed on a CRT screen. Using the zoom function enables one to fill the screen with a specified portion of the drawing detail. This is a viewing facility that could be compared to having a powerful magnifying glass to examine fine details. The zoom activity does not modify the drawing in any way. Figure 11-11 shows a window/section of a drawing to reveal fine detail.

Objects that are symmetrical about an axis may be *mirrored*; that is, the operator may create one-half of a view of an object on the CRT and then automatically produce the second half. Figure 11-12 illustrates this feature.

The **rotate** function enables the user to turn a two- or three-dimensional part displayed on a CRT through a selected angle about an axis. Dif-

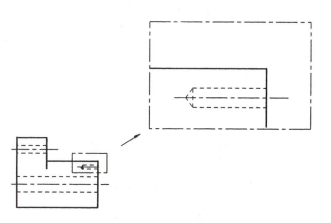

FIGURE 11-11
Zoom function.

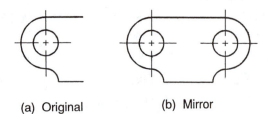

(a) Original   (b) Mirror

FIGURE 11-12
Mirror function.

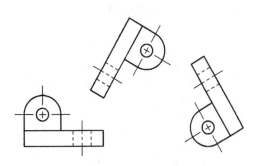

FIGURE 11-13
Rotate function.

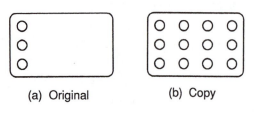

(a) Original   (b) Copy

FIGURE 11-14
Copy function.

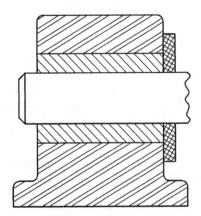

FIGURE 11-15
Crosshatching.

ferent views of the object as illustrated in Figure 11-13 can be automatically created on the screen. The designer or drafter can select the most appropriate one.

A CAD system also has the ability to **move** a portion of the display that has been positioned incorrectly on the screen. This function is accomplished by creating a window around the desired information and moving it to the new location by means of the cursor or by using coordinates.

The **copy** function saves the drafter considerable time when there are a number of objects or symbols that need to be repeated. Once the original has been created, the copies are positioned by the cursor or by coordinates. Figure 11-14 illustrates the copying of three sets of additional holes for the object. The copy function is similar to the move function.

It is often necessary to produce cross sections on an engineering drawing to reveal the interior details of a certain component. Cross sections are commonly indicated on a drawing by **crosshatching,** which if done manually is very time consuming and tedious. Most CAD systems can do this task automatically. The appearance of crosshatching varies depending on the hatch angle, the spacing between successive hatch lines, and the pattern. Figure 11-15 illustrates this variation in cross-hatching.

Most popular CAD systems provide a layering capability in which different drawing attributes can be assigned to different levels or **layers** within a drawing. For instance, separate layers can be created for construction lines, dimensions, drawing notes, hatching, etc. Layers can be turned on and off for display and printing, frozen or unfrozen so accidental changes are made, and assigned different colors. Using layers can significantly help a designer to organize a drawing.

Most CAD systems provide the operator with a complete range of automatic dimensioning conventions that cover the ANSI standards. The capability to create dimensions to a standard format is a prime requisite of a good CAD system. Dimensioning of a drawing on a CRT screen is done automatically or by the operator using a menu. In either case, extension lines, dimension lines, and arrowheads are supplied in accordance with good practice. Automatic dimensioning is possible because the object is defined in space by its XYZ coordinates.

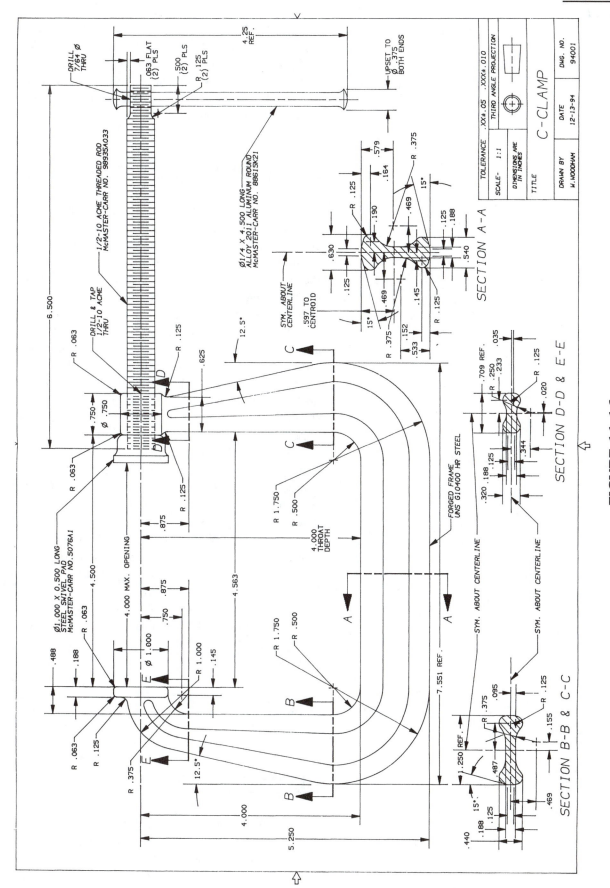

FIGURE 11-16
Complete drawing of a deep-throat
C-clamp.

The endproduct for any CAD system and designer is a completely dimensioned drawing of a component or system that is ready for production. Figure 11-16 shows a deep-throat C-clamp that is ready for production. Note the complete dimensions, crosshatching, drawing notes, etc.

ADVANTAGES

All the enhancements given in the preceding section are also the advantages that CAD enjoys when compared to instrument drawings. In addition to those advantages, and to some extent because of them, CAD is faster, more accurate, more consistent, neater, easier to store, and easier to make corrections and revisions.

Once a CAD operator is trained, lettering and linework tasks are much faster than instrument drawings. CAD hardware is accurate to 0.001 inch, and this cannot be duplicated by the human drafter. Linework and lettering are the major strengths of CAD because the output is neat and consistent.

One of the major advantages of CAD is better storage capability. Computer storage of drawings is much more space efficient than manual storage. Also, the deterioration problem associated with traditional drafting media is eliminated.

An additional advantage of CAD occurs when a drawing needs to be corrected or revised. All the advantages of creating a drawing are applicable to this activity.

A major advantage is the link with manufacturing. The information stored in the computer can be used to manufacture the part or assist in its manufacture.

CAD is one of the foundations of a paperless company. Access to any drawing is accomplished by direct linkage, by telephone lines, or by satellite to the database.

GEOMETRIC MODELING

Geometric modeling software is often referred to as 3D modeling, solid modeling, rendering, and light source shading, to name a few. A three-dimensional model allows the designer to inspect a

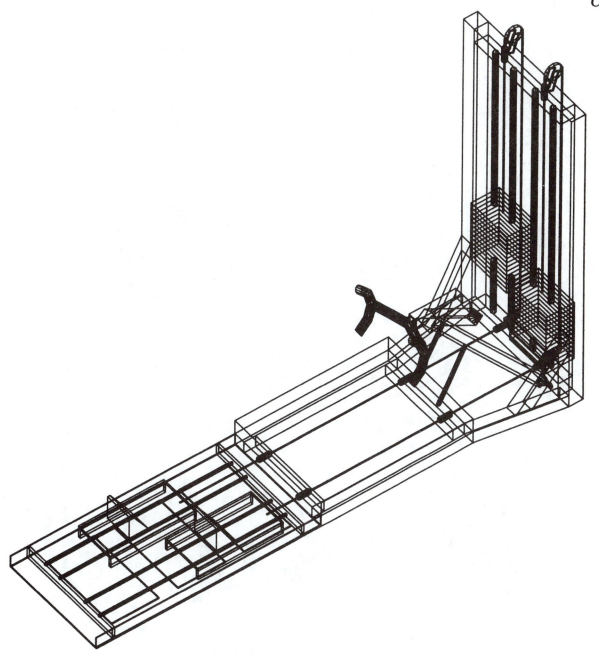

FIGURE 11-17
Three-dimensional drawing
of an exercise machine.

design visually and identify potential problems prior to analysis and production. As shown in Figure 11-17, the overall workings of an exercise machine and interactions between components can easily be seen. In Figure 11-18, the launch pad arrangement for the space shuttle is depicted in two planar views and a third view, which is three-

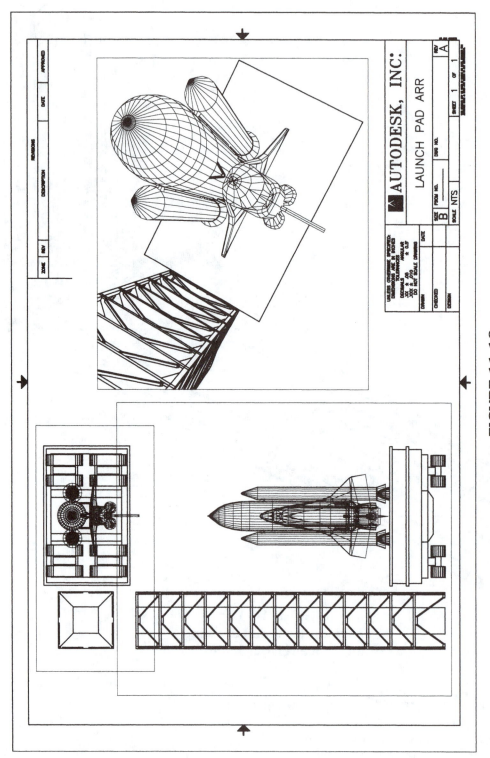

FIGURE 11-18
Planar and three-dimensional drawing
of the launch pad arrangement
for the space shuttle.

FIGURE 11-19
Shaded drawing of deep-throat C-clamp.

dimensional. The three-dimensional view helps a designer interpret the planar views, which might not be quickly understandable at first glance or without the help of the three-dimensional view.

To further help the designer, many geometric modeling software packages have the ability to render or light source shade components. In Figure 11-19, the deep-throat C-clamp depicted in Figure 11-16, has been shaded with gray scales. A rendered or shaded drawing can be very beneficial to a company. For example, executives can easily make decisions concerning potential sales, the sales force can show a rendered drawing to customers prior to production, and production workers can visualize the final product before proceeding with production, to name a few. Numerous geometric modeling software packages are available; however, most CAD packages have geometric modeling built into the packages or can be added on.

SUMMARY

In summary, one must know or do the following:

1. The term CAD/CAM stands for "computer-aided design and computer-aided manufacturing."

2. A computer system consists of three main areas of data handling—input, processing, and output—and is backed by a fourth, storage.

3. Computers are categorized as micro, mini, and mainframe.

4. A typical interactive graphics workstation consists of a terminal, input devices, and output device.

5. CAD hardware consists of CRT, keyboard, digitizer/puck or light pen, mouse, plotter, and printer.

6. The three main categories for CAD software are operating, applications, and graphics.

7. Input methods are digitizer/puck or light pen, mouse, and keyboard. These methods are used in conjunction with a menu.

8. CAD is enhanced by the following functions: zoom, mirror, rotate, move, copy, crosshatch, automatic dimension, and spatial coordinates.

9. In addition to the enhancement advantages of CAD, it is faster, more accurate, more consistent, neater, easier to store, and easier to correct or revise than instrument drawing. Also, the link with manufacturing is easily effected.

10. Geometric modeling allows the designer to inspect a design visually and to identify potential problems prior to production and use.

DRAWING IN CAD

CHAPTER

12

Upon completion of this chapter, the
student is expected to:

- Know how to use the drop-down menus.

- Be able to use the AutoCAD toolbox.

- Be able to set up drawing layers.

- Complete a drawing.

- Send a drawing to a plotter or printer.

INTRODUCTION

One of the components of a computer-based drawing system is the software or program that allows you to do the graphics and to print the results of your work. Two of the more popular programs are AutoCAD and CadKey. The AutoCAD system has a student version of its software called AutoCAD LT. This software has the capabilities of the professional version for drawing in two dimensions. It can also work for making isometric drawings. AutoCAD LT does not have three-dimensional, wire frame, or solid modeling capabilities. This chapter will address the basic concepts necessary to begin drawing in the AutoCAD LT Windows version.

Designing and drawing on a computer involves being able to communicate with the CAD program you are using. Communicating with CAD is much like learning a new language. The more you use it the stronger your vocabulary becomes, making it easier to produce clearly understood drawings. You must tell the computer, through a series of commands, where the lines, arcs, circles, dimensions, etc., are to be placed on the drawing.

It as assumed that the reader is familiar with the basic requirements for working with Windows and Windows applications. If this is not the case you will need to learn the Windows system before you can understand the concepts of AutoCAD LT.

THE GRAPHICS WINDOW

The first step in learning to communicate with AutoCAD LT is to know where to find the tools you need. Figure 12-1 shows the major elements of the graphics window. The **title bar** at the top of the screen displays the name of the drawing file that is being used. The control menu on the left and the window sizing buttons on the right are also a part of the title bar.

Directly below the title bar is the **pull-down menu bar.** Pull-down menus can be accessed with the mouse or keyboard. The pull-down menus are the most comprehensive set of commands available to the AutoCAD LT user.

Next is the **toolbar.** The buttons on the tool bar provide mouse access to frequently used drawings settings including file commands, plot or print-

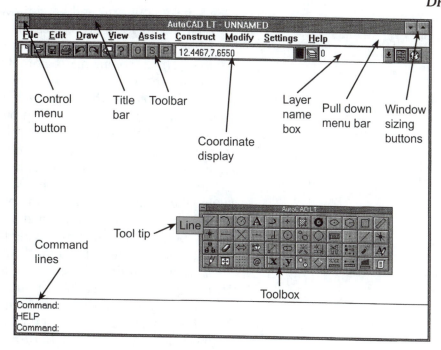

FIGURE 12-1
Graphics Window

ing commands, and layer controls. Information about the drawing such as the X and Y location of the cursor and current drawing layer are also displayed on the tool bar.

The **tool box** is a series of buttons, referred to as *tools,* in a floating window that can be moved as necessary to keep it from interfering with the working area of the window. This toolbox is the most direct method for selecting commands with the mouse. When the cursor points at any tool a small tag pops up. The tag is called a **tool tip,** and tells the user the name of the tool being selected.

Command lines appear at the bottom of the screen. The command line is the central point of communication with the AutoCAD LT program. It takes all the input from the keyboard or mouse, lists the command options available, and asks for input required to carry out commands.

ISSUING COMMANDS

Issuing of commands is one operation that is fundamental to all technical drawing. When drawing by hand your brain issues the commands necessary to make your arm and hand move in the required direction to form lines, arcs, or circles. Similarly, when drawing in a CAD system you have to give the correct commands before the program can make a move.

Most commands follow a pattern in their operation. You issue a command by picking a drawing tool, such as the line tool. The command line will prompt you for a starting point. The command line will read "_Line From Point:" as shown in Figure 12-2. You can respond by clicking and releasing the left mouse button at the point you want the line to start or enter an X and Y coordinate, such as .500, .500, on the keyboard. The command line will next prompt you for the next point for the line. You can drag the line with the mouse to the next point and click the left button to anchor the line or enter the coordinates from the keyboard. In either case, when you input your requirements you must press the *Enter* or *Return* key for the command to be completed. When the line is complete, pressing the enter key a second time will end the command. It is important to remember to watch the command line for the prompts or input requests the computer needs from you to complete a command. Many of the commands such as the line

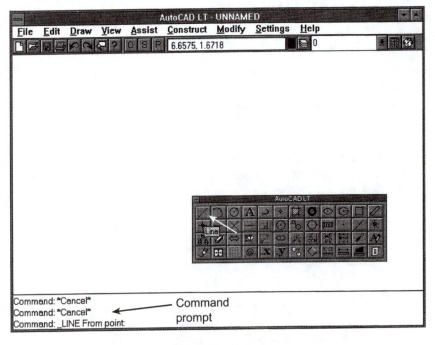

FIGURE 12-2
Command line prompt

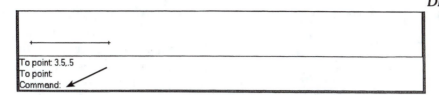

FIGURE 12-3
New command prompt

command continue until you end them. Others, such as the arc or circle commands, end when the object is drawn. When the command has been completed the command line will show the word "Command" on the bottom line as shown in Figure 12-3. If you want to repeat a command that was just used, press the enter key and respond to the command prompts to make any changes to the size or location of the object you want to draw.

Commands can be canceled at anytime by pressing the *Control* key $\boxed{\text{CTRL}}$ and the letter *C* key on the keyboard at the same time. This is a very important command to remember when you are starting to learn the CAD system because it is easy to get confused and not want to complete a command you are unsure of using. The "Undo" command and "Undo" button on the tool bar will reverse a command just as in any Windows application.

THE PULL-DOWN MENUS

The pull-down menus are the most comprehensive set of commands available to AutoCAD LT users. A complete description of all the available commands would be more confusing than helpful to the beginning user of the program. This section will highlight only the areas that will be useful in being able to produce basic drawings.

The file menu (Figure 12-4), is used to start a new drawing, open an existing drawing, save your work, print a hard copy of drawing, and exit the AutoCAD LT program.

Saving your work often will keep you from having to redraw what was already done should there be a power, hardware, or software failure. Work can be saved automatically by opening the "Preferences" dialog box in the file menu and selecting the "Auto Save" check box. The length of time between saves can also be set in this dialog box (see Figure 12-5).

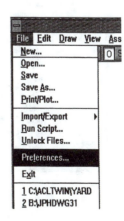

FIGURE 12-4
File menu

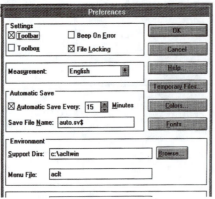

FIGURE 12-5
Preferences dialog box

The edit menu allows you to undo a previous command or redo a command that was accidentally erased with the undo command. These commands can also be executed by clicking the mouse on the appropriate tool bar buttons.

The draw menu (Fig. 12-6), is used to give the drawing commands that are required of every drawing. These commands include drawing lines, arcs, and circles. There are submenus for some command lines. They have an arrow at the end of the command line. The "Arc" command, for example, gives you a choice of points to be used for the location of the arc. Additionally, the draw menu is used for adding text, hatch lines, and dimensions to your drawing. Many of these commands can be made by clicking the corresponding buttons in the tool box.

The view menu (Fig. 12-7), has a "Zoom" command that allows you to enlarge an area of the drawing. Working on small or very detailed areas of a drawing can be easier when they are blown up or enlarged. The drawing can also be redrawn through the view menu. This "Redraw" command will remove any phantom lines or blips that may be

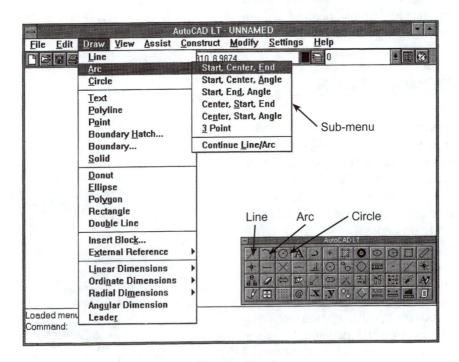

FIGURE 12-6
The draw menu and corresponding tool box

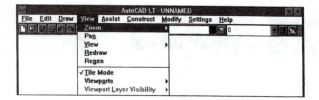

FIGURE 12-7
The view menu

left on the screen from erasing, canceling commands, or using the "Undo" command.

The construct menu can be very helpful in the duplication of parts of a drawing. The "Array" command will allow you to produce multiple copies of objects such as circles in a specific rectangular or radial pattern. "Copy" can be used to duplicate an object and place the copy at a specific location within the drawing. "Mirror" creates a new version of an object such that it becomes a reflection of the original in respect to a predetermined line or plane as shown in Figure 12-8. The "Chamfer" and "Fillet" commands are used to apply angles or radii to intersecting surfaces as shown in Figure 12-9.

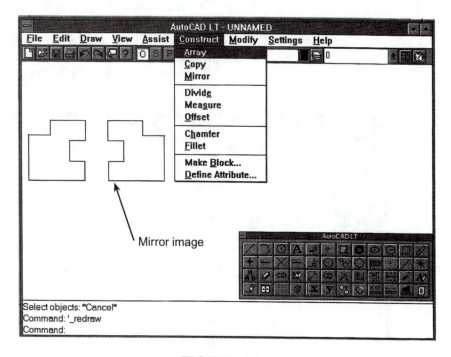

FIGURE 12-8
Mirror command

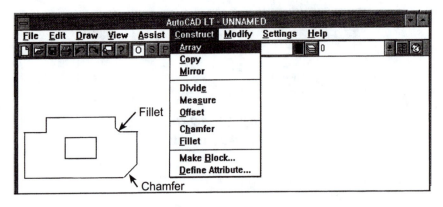

FIGURE 12-9
Chamfers and fillets

The modify menu (Fig. 12-10) is used to make changes to objects that have already been drawn. The "Erase" command is used to delete an entire object or line. Anything erased unintentionally can be restored by clicking the mouse on the "Oops" command. This command can also be used to restore an object that was lost with the "Undo" command. Also under the modify menu are the commands for moving ("Move"), turning ("Rotate"), lengthening ("Stretch"), and shortening ("Trim") them. The "Modify" command is used to edit or change text, section or hatch lines, and dimensions. Many of the commands found in the modify menu are also available in the tool box. The help menu is used in the same manner as in

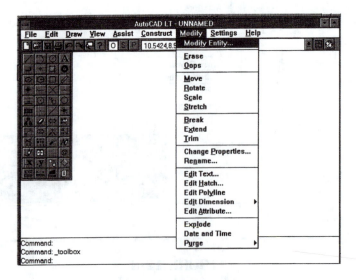

FIGURE 12-10
The "Modify" menu

all Windows applications. It may be accessed by using the F1 button or clicking on the help menu.

THE TOOL BOX

Many of the commands required to produce a drawing can be issued by clicking on the appropriate button in the tool box. As you point the mouse to the button the tool tip will appear with the name of the tool in it. Tools may be used in combination with other tools to give a complete set of commands. For example, if you had two horizontal lines and needed to draw an angular line from the left end of the upper line to the right end of the lower line you would use the line tool button and the end point button. The first move would be to click on the line button. The command line would prompt for a starting point with the words "from point." Clicking on the end point button and then on the upper line near the left end will answer the prompt for the starting point. The next prompt will be for the second point, or in this case the ending point, of the line. By clicking on the end point button again and then picking a point near the right side of the lower line and clicking the mouse button, the line will be drawn from the left end of the upper line to the right end of the lower line. You must press the "Enter" key on the keyboard to confirm the line command.

The most frequently used tool buttons are:

Line	Arc	Circle
Text	Ellipse	Polygon
Rectangle	End point	Intersection
Midpoint	Perpendicular	Center
Tangent	Quadrant	Nearest
Copy	Erase	Move
Extend	Trim	Edit text
Redraw		

LAYERS AND LINETYPES

Working in layers is like drawing on transparencies. You might use one transparency for the object views, another for the dimensions, another for text, etc. The transparencies could then be laid on top of each other to form a complete drawing or viewed one at a time to reduce the complexity of a very detailed drawing. Working in layers also provides a method for using different linetypes for objects. You need solid lines, centerlines, and dashed lines for many drawings; each type of line can be assigned to a different layer.

When you start a new drawing the only line type available is the continuous line. You must load the other line types before you can use them. To load the linetypes click on the settings button on the drop-down menu bar. Then click on the "Linetype style" and "Load" lines (see Fig. 12-11). On the command line, enter an asterisk (*), and press the "Enter" key. Another dialog box will open as shown in Figure 12-12. Click on the "Aclt.lin" name and click the "OK" button. The lines will be loaded at this time. Press the "Enter" key once more to confirm the command. Now that the linetypes are loaded you can assign a different type of line to different layers of the drawing. To create a layer from the toolbar, click on the layer control dialog box as in Figure 12-13; the layer control dialog box (Fig. 12-14), will open.

If you want to create a layer for only center lines, move your cursor to the text box near the bottom and enter a name, such as center, then

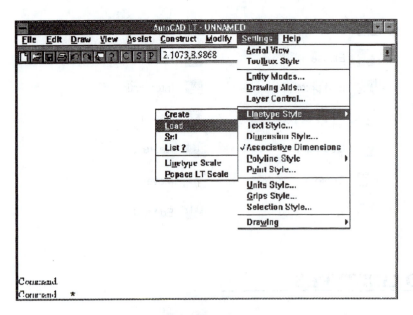

FIGURE 12-11
Loading linetypes.

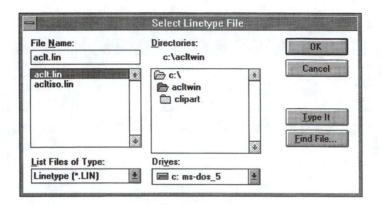

FIGURE 12-12
Linetype files

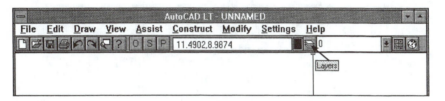

FIGURE 12-13
Layer control dialog button.

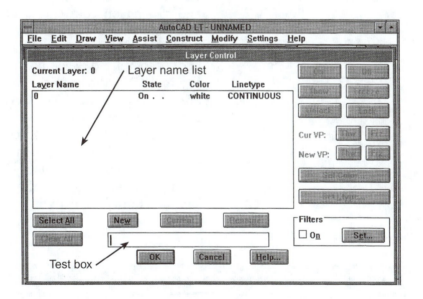

FIGURE 12-14
Layer control dialog box

click on the "New" button. The layer that has just been created will now appear in the layer name list. You can enter as many layer names as you want at one time by separating each name with a comma and no spaces. When the layers are

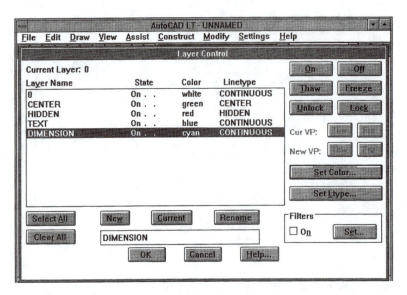

FIGURE 12-15
Layer control dialog box.

named, each can be given a different color and linetype simply by clicking on the layer you want and clicking the "Set Color" or "Set Ltype" buttons and clicking on your choice of color or linetype. The word "Continuous" under "Linetype" means a solid line and would be used as a visible or object line. A completed dialog box might look similar to Figure 12-15. When all the lines you want to use are set in layers, click the "OK" button at the bottom of the dialog box.

After the linetypes and colors have been assigned to each layer you can make any layer current by clicking on the layer control box (Fig. 12-16) on the toolbar and highlight the layer name for the layer you want to be current.

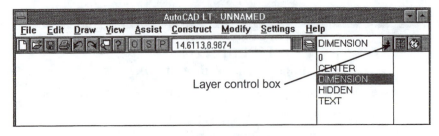

FIGURE 12-16
Current layer box.

DRAWING IN AUTOCAD LT

It was stated earlier that the best way to learn the language of drawing in CAD is to use it. The remainder of this chapter will lead you through the steps necessary to make and print a simple drawing. Figure 12-17 is a sketch of the drawing that will be made.

Open AutoCAD LT from the Windows Program Manager by double-clicking on the AutoCAD LT icon. If the "Create New Drawing" box (Fig. 12-18) does not appear when the program opens, choose "New" from the file menu on the toolbar. In the "Create New Drawing" box under "Setup Method" click on "Quick" and then click the "OK" button. The "Quick Drawing Setup" dialog box will open. "Decimal" should be displayed in the "Units of Measurement" box and the "World [actual] Size to Represent" area should show the default settings of 12.000 in the width box and 9.000 in the height box. If your screen does not reflect these settings, click the appropriate box and change the settings to look like Figure 12-19.

The "Settings in World" represent the size of the area you will need in order to make your draw-

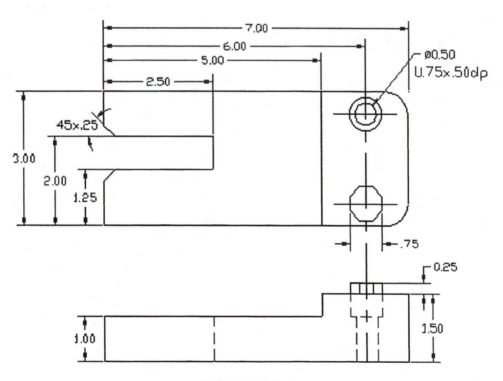

FIGURE 12-17
Sketch of a simple drawing

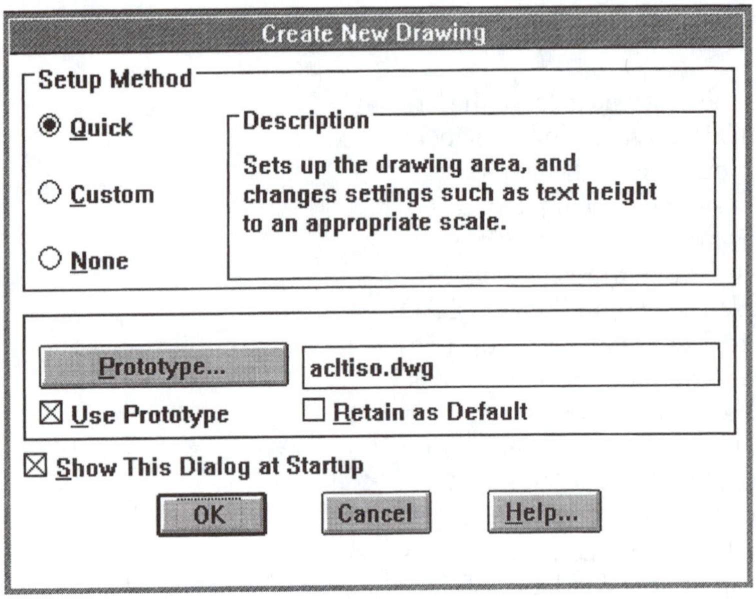

FIGURE 12-18
Create new drawing dialog box.

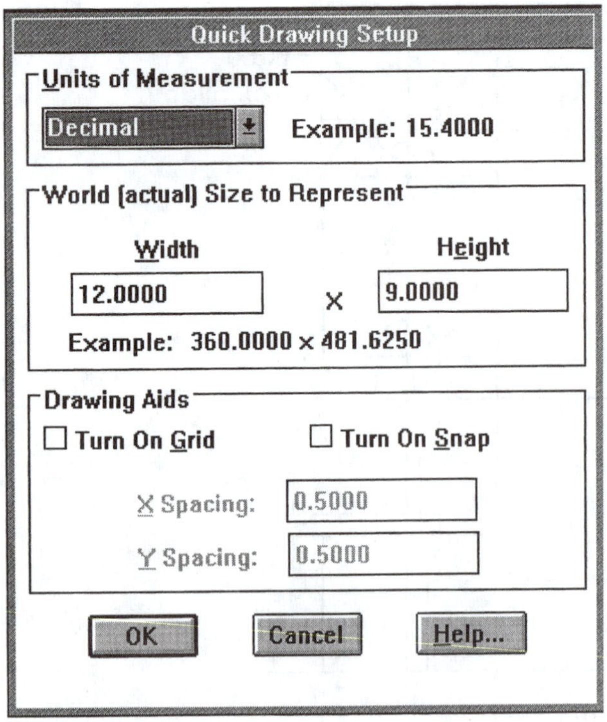

FIGURE 12-19
Quick drawing setup dialog box

ing to full scale and leave an invisible border around it. Since our drawing will be 7 inches wide the 12-inch default setting will leave room for dimensions and the invisible border. The height of the two views will equal 4½ inches and space needs to be left above, below, and between the views for dimensions. A height of 9 inches will meet all the requirements. "Grid" and "Snap" modes are tools that will aid you in aligning your drawing objects. These modes are set in the "Drawing Aids" box of the "Quick Set Drawing Setup" dialog box. For this drawing click on both the "Turn on Grid" and the "Turn On Snap" boxes. Click the mouse in the "X Spacing" box and change the dimension to 0.250. Next click on the "Y Spacing" box. You will see that the program will automatically update that box with the same value as the "X Spacing" box. When you complete the changes click on the "OK" button. You should have a graphics area with dots indicating the .250 grid spacing on your display.

Next, you will set the precision of the units to two decimal places. From the settings menu click the unit style. In the "Units Control" box click the down arrow under precision and scroll through the list until you see a two place (0.00) decimal. Highlight that setting and click "OK."

Another drawing tool is the Ortho mode. When the Ortho mode is turned on it restricts the cursor to move only in horizontal and vertical directions, making it easier to draw straight lines. To turn the Ortho mode on click the "Ortho" button ⊡ on the toolbar. The Ortho mode button is a toggle. Clicking it in the "On" position will switch the mode to off and allow you to draw lines on an angle. Since our drawing will require using hidden lines and centerlines, it is necessary to load the linetypes for these functions. Load the Linetype now by following the steps outlined in the previous section.

Drawings contain many different components. These components can be grouped by creating a different layer for each group of components. We will create four layers for our drawing:

CENTER   for drawing centerline

DIMS   for dimensions

HIDDEN   for hidden lines

OBJECT   for the visible part lines

From the settings menu click "Layer Control." Clicking the "Layer Control" button on the toolbar will also open this dialog box. Move the cursor to the text box near the bottom of the dialog box and type the word "center"; next click the "New" button. The layer you created will appear in the "Layer Name" list. Move back to the text box and this time type the word "dims" and follow it with a comma [,] continue with the word "hidden" and a comma and then the word "object." Click the "New" button and you will see the names added to the layer name box. To help distinguish different lines you can assign specific linetypes and colors to each layer. Use the mouse to highlight the layer named "Center." Now choose the "Set Color" button on the right side of the layer control dialog box. A color palette will open up. Click on the color red and click the "OK" button. The Center layer will now have red under the color column. Next choose the "Set Linetype." Another dialog box opens showing the various linetypes available. Click on the linestyle named "center" and click "OK". The Center layer will now have the word "CENTER" in the linetype column. To change color and linetypes for the other layers you must first click on the center layer to complete the commands to that line and then click on the next layer name you want.

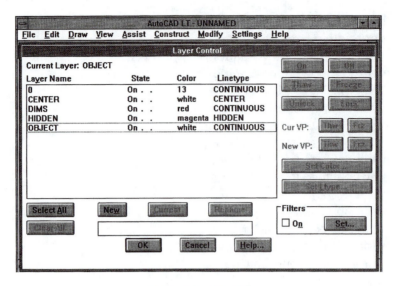

FIGURE 12-20
Layer control chart

When finished, the layer control chart should look similar to Figure 12-20.

To make a layer current click the down arrow on the layer control box on the toolbar and select the appropriate layer. When a layer is current everything you draw is on that layer until you change layers.

The settings are now complete and you are ready to start drawing. Before you continue, save the work you have completed. To save the drawing choose the file menu and click on "Save As." Enter the name "SUPPORT" in the File Name box, or choose a name you prefer, and click on "OK." Saving your work often ensures that it is not lost in the event of a power failure or other unforeseeable incident.

Many different methods can be used to draw objects. As you become familiar with the workings of AutoCAD LT you will develop methods that you are most comfortable with. The approach used for this drawing will use a variety of different tools and commands to help you become more familiar with them.

To draw the top view you will want the Ortho mode to be turned on. If it is not on, click the "Ortho" button on the tool bar. Also, check the layer box to be sure you are starting your drawing in the "OBJECT" layer.

To draw the top horizontal line of Figure 12-17 choose "Line" from the toolbox or draw menu. Click and release the left mouse button at a point about 2 inches from the top of the drawing area and 2 inches to the right of the left edge of the drawing grid dots. This action will anchor the starting point of the line. Drag the cursor to the right until you are approximately 2 inches from the right edge of the grid. Click and release the left mouse button to anchor the ending point of the line. Press the "Enter" key to confirm the command.

The next step will be to use the offset command to copy and locate the other horizontal lines needed for the top view. From the construct menu choose "Offset." On the command line enter "1.00" for the offset distance and press the "Enter" key. The cursor will change from cross hairs to a square box. This box is called a **pickbox.** The pickbox is used to select objects you want to copy, move, or erase. The command line will now ask you to select the object you want to offset. Click on the line you just drew. Now the command line asks for the side of the selected line on which you want to place the new line. Click the mouse anywhere below the line you selected and a second line will be drawn 1 inch below the first. Press the "Enter" key to confirm the command. This line will be the top line of the slot. At this time we are not concerned about the length of the line because it is very easy to erase parts of a line at a future time.

Draw the two remaining horizontal lines of the top view by using the offset command in the same manner as you did in the last command. Remember that you will need to change the offset distance for each line; your drawing should look like Figure 12-21. There are small marks on the screen in places where you clicked to specify points to offset the lines. These marks are called blips. A blip is a reference point that indicates where you clicked the mouse. The redraw command will remove these blips from your drawing. From the toolbox or the view menu choose "Redraw." Pressing the F7 key twice will also redraw the view.

Next we will draw the vertical lines. To draw the vertical line on the left side choose "Line" from the toolbox; next choose "Endpoint" from the toolbox. Move the cross hair cursor to a point near the left edge of the top line, click and release the left mouse button to anchor the line at the edge of the top line. By using the endpoint tool you do not have to be exactly at the end of the line when you click the mouse; the program will place the line on the end of the line you have specified. Click the

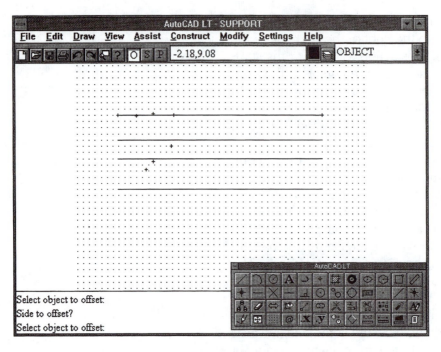

FIGURE 12-21
Horizontal lines of the top view

"Endpoint" tool once more and drag the cursor to the lower line of the top view and click the mouse to anchor the line at the end of the lower line. Press the "Enter" key to confirm the command.

Use the "Offset" command to copy the line to the locations needed for the remaining three vertical lines. The offset distance for the first command will be 2.50 and you will need to specify the side to offset as the right side of the existing line. Your drawing will look like Figure 12-22.

All the vertical and horizontal object lines are now drawn for the top view. They need to be trimmed and some areas may need to be erased in order to make the object take the required shape. From the toolbox click on the "Trim" button. The command line will ask you to select the cutting edges. Cutting edges are the stopping points for the lines you want to trim or erase. Click on the far right vertical line and press the "Enter" key. The next command line will ask for the objects to be trimmed. Click on each of the horizontal lines in the area to the right of the vertical line you chose as the cutting edge. Each of the lines should be erased as you click on them. Press the "Enter" key to confirm the command. Press "Enter" once more to reactivate the trim command and choose a new cutting edge. This time, click on the second vertical

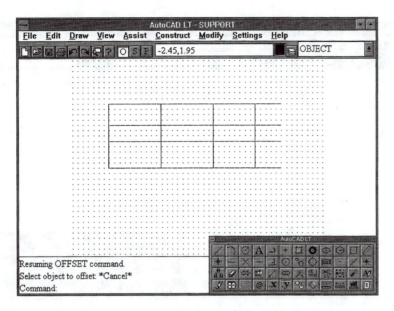

FIGURE 12-22
Vertical and horizontal lines for top view.

line from the left. Press the "Enter" key and then click on the second and third horizontal lines at a point that is to the right of the cutting edge. These two lines should be erased back to the second vertical line. Press the "Enter" key again to confirm the command. Continue this process to remove the vertical lines above and below the slot and the vertical line that will open the slot on the left end of the object. Figure 12-23 represents what your drawing should look like at this time.

To draw the chamfers (angles) on the front edge of the slot choose the "Chamfer" command from the construct menu. In the command line enter the letter "D" to set the distance of the chamfer; then press the "Enter" key. Next enter ".25" for the first distance and press the "Enter" key. Enter ".25" again for the second distance and press "Enter." Press "Enter" again and click on the upper left vertical line and then click the upper horizontal line of the slot. The intersection of those two lines should change to an angular line. Press the "Enter" key twice to end the command and restart the same command to draw the second chamfer. Repeat the process using the lower horizontal line of the slot and the lower left vertical line.

The other end of the object needs to have a radius on each corner. These radii can be added by using the fillet command in the construct menu. After selecting the fillet command, type in the

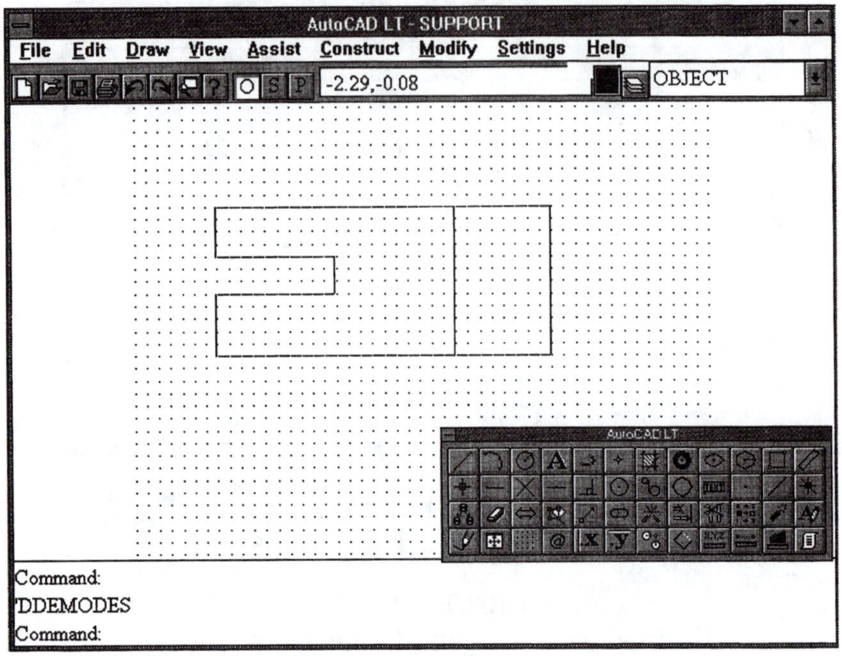

FIGURE 12-23
Trimmed lines.

letter "R" for radius and press the "Enter" key. The command line will then request a radius dimension and show a default dimension of .500. Since that is the size of the radius you want, you can press the "Enter" key twice to confirm the dimension and open the command for selecting the lines required to form the radius. Now you will click on the upper horizontal line near the right side and then on the far right vertical line. The intersection of those two lines will form the radius. The radius command is completed without having to press the "Enter" key. Press the "Enter" key to repeat the line selection process using the lower horizontal line and the right vertical line. The drawing should look like Figure 12-24 at this time.

Click the arrow on the drop-down layer control box on the toolbar and click on the layer named "CENTER." The lines that are drawn next will have the center line format and color you selected for your centerline layer. You need a vertical centerline 1 inch to the left of the right end of the object. Since we know that the grid in the drawing area is set at .250 of an inch we can find the 1-inch mark by counting over four dots on the grid. Click the button on the toolbar with the "S." This sets the draw mode to snap, which means the com-

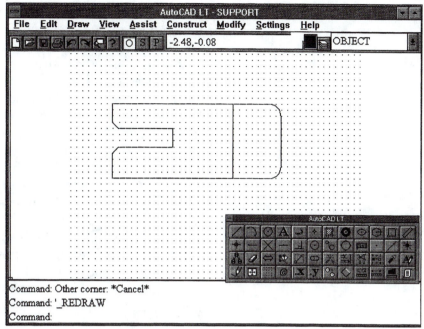

FIGURE 12-24
Chamfer and fillet lines.

mands will be executed only at .250 of an inch intervals. Click on the "Line" button in the toolbox and click and release the left mouse button on the top horizontal line and the fourth grid from the right vertical line. Drag the cursor straight down to the lower horizontal line, and click and release the mouse to anchor the line. Press "Enter" to confirm the command. The line should be the color and style of a centerline. The top horizontal center line can be located by coming down two grid lines from the top line and the second horizontal center line can be placed by using the offset command. After the centerlines are drawn, click on the "Snap" button on the toolbar to turn the snap command off and change the layer control box back to the object layer.

To draw the circles for the counterbored hole select the "Circle" button in the toolbox. The command line will have center point as the default setting. Click the "Intersection" button in the toolbox and click on the point where the upper centerline crosses the vertical centerline. The command line will change to a default of radius and a default radius size. Type the letter "D" to change the default to diameter and press the "Enter" key. Next enter the dimension .38 and press the "Enter" key to end the command. A circle should appear in the upper right area matching the sketch in Figure 12-17. Repeat this procedure using the .75 diameter for the counterbore size.

Drawing the octagon knob is similar to drawing the circles. Click the "Polygon" button in the toolbox and then click the "Intersection" button in the toolbox. Click the cursor on the intersection point of the lower horizontal centerline and the vertical centerline. Type in ".38" for the radius of the circle that the octagon sides will touch and click "Enter" to confirm the command. This completes the lines of the top view.

Before starting the front view move the toolbox to the right hand anchor position. Choose the settings menu and then click on the toolbox style command. Repeat these moves until the toolbox appears on the right side of the screen.

To begin the front view select the line tool from the toolbox. Click the cursor on a point about 1 inch up from the bottom of the drawing area and to the left of the top view. Drag the cursor to the right until it is slightly beyond the right side of the top view and click the mouse. Press the "Enter" key to confirm the command. This will be the lower horizontal line of the front view. Draw the three remaining horizontal lines by using the offset command. The drawing should look like Figure 12-25.

Drawing the vertical lines can be accomplished by extending the vertical lines from the top view to the front view. From the toolbox click on the "Extend" button. Next click the lowest horizontal line on the drawing and press "Enter" to set the end point of the extended lines. Now click the cursor on each of the vertical lines of the top view. The lines should extend down to the bottom horizontal line of the front view. Press the "Enter" key to confirm the command. To avoid the confusion of having too many lines crossing each other, use the trim command to remove as many of the lines as possible. Start by clicking the "Trim" button in the toolbox. Click the bottom line of the top view and the second line from the bottom of the front view and then press the "Enter" key. These moves set the points between which the line chosen in the next command will be trimmed. Click on the left vertical line at a point between the two previously chosen lines to erase that portion of the line. Press the "Enter" key twice to confirm the command and begin a new trim command. This

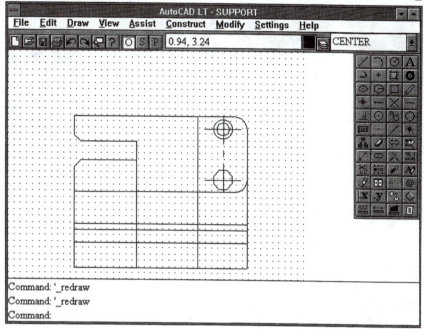

FIGURE 12-25
Counterbored hole, octagon knob,
and front view construction.

time click the cursor on the lower horizontal line of the slot and again on the second line from the bottom of the front view and press the "Enter" key. Click the vertical line that extended the back of the slot to the front view at a point between the two views to erase it. When you have completed trimming all the lines possible at this time, the drawing should look like Figure 12-26.

It will be easier to draw the front view of the octagon if the area is enlarged. To enlarge an area choose the view menu and click on the "Zoom" command and then click on "Window." Move the cursor to a point just above and to the left of the octagon in the top view and click the mouse. Drag the cursor to a point below and to the right of the front view and click the mouse.

Click on the "Line" and "Intersection" buttons in the toolbox and then click and release the cursor on the left side line of the octagon below the horizontal center line in the top view. Drag the cursor down to the third line from the bottom of the front view, and click and release the mouse to anchor the line. Press the "Enter" key to confirm the command. Repeat this operation for the other three lines of the octagon front view. The magnified view

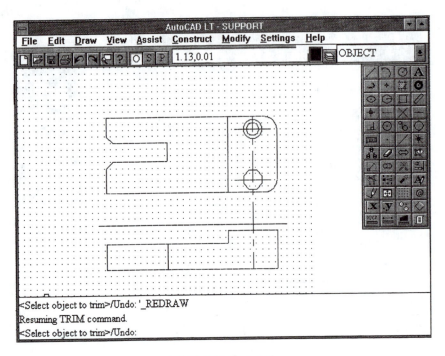

FIGURE 12-26
Front view with trimmed horizontal lines.

will look like Figure 12-27. Trim the lines between the two views and the top horizontal line of the front view before returning the drawing to its original size. To return to the full-size view select the view menu and click on the "Zoom" command

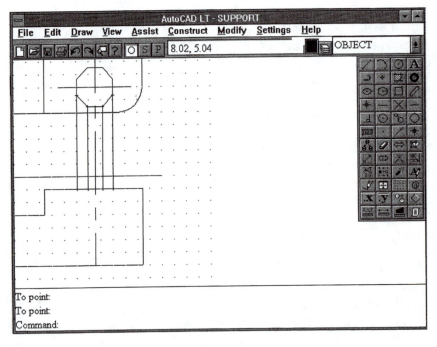

FIGURE 12-27
Magnified view of the octagon knob.

and then click on the "All" command. The drawing will return to its original size.

Change the layer control box to the "HIDDEN" layer and draw the lines for the holes in the front view. Click on the "Line" button and the "Quadrant" button of the toolbox. Click the cursor on the lower left side of the larger circle in the top view. Drag the line down to the lower horizontal line of the front view and click the mouse to anchor the line. Repeat the process for the other three hidden lines required for both holes. Trim the lines between the top and front views. To ease drawing the horizontal line for the counterbore use the zoom command to enlarge the area. Next click the line tool and click the cursor on the second grid line ½ inch below the top surface of the front view and intersecting the leftmost hidden line. Drag the cursor across to the leftmost hidden line and click the mouse to anchor the line. Press "Enter" to confirm the command.

The final step to complete drawing the objects is to change the line in the front view that represents the right edge of the slot to a hidden line. In the modify menu choose "Change Properties." Click the cursor on the second vertical line from the right in the front view. The "Change Properties" dialog box will open. Click on the layer bottom and on the hidden name in the layer control box then click "OK" to close the layer control box and "OK" again to close the "Change Properties" box. The line will change to the color and configuration of a hidden line. The drawing is ready to be dimensioned and should look like Figure 12-28.

Before starting to add the dimensions change the layer control box to the "DIMS" layer. Adding dimensions to the drawing is accomplished through the draw menu. From this menu choose linear dimensions and then horizontal. The command line will ask for the location of the first extension line. That is the line on the drawing that you will use to start this dimension. In the toolbox click the "Intersection" button and then click the mouse on the intersection of the upper left corner of the top view. The command line then needs the second extension line point. Again click on the "Intersection" button and click the mouse on the point where the upper line and the right vertical line of the slot come together. The command line will ask for the dimension line location. Move the mouse to the area slightly above the top of the top

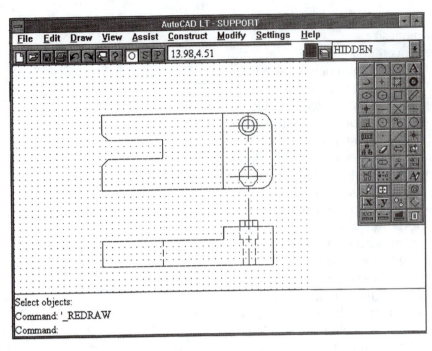

FIGURE 12-28
Completed top and front view shapes.

view and click the mouse. This move will set the location of the dimension line. Next the command line will ask for the dimension text and show a default of the dimension that the program measured between the lines picked for the dimension. If the object was drawn accurately, this default dimension will be accurate. If not, type in the dimension value you want and press the "Enter" key. Continue adding the remaining horizontal dimensions in the same manner. The vertical dimensions are completed by selecting the "Vertical" command from the draw menu. To dimension the chamfer choose the angular command from the draw menu and pick either the vertical or horizontal line that intersects with the angular line for the first line and the angled line for the second line; then place the cursor in the location for the text to appear. Dimension text can be edited by choosing the modify menu and selecting the "Edit" dimension submenu. The completely dimensioned drawing should be similar to Figure 12-29.

The final step is to plot or print a hard copy of the drawing. From the file menu choose "Plot." A plot configuration dialog box will open. Click on the "Rotation and Origin" bar and another dialog box will open. This box allows you to configure the orientation of the printing. In the "Plot Rotation"

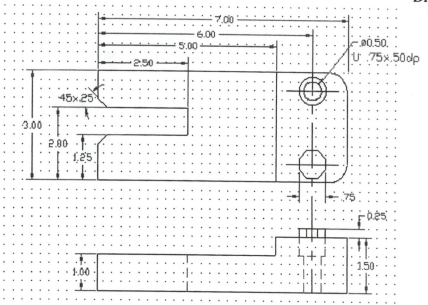

FIGURE 12-29
Completed drawing.

dialog box click on "90" to have your drawing print in a landscape mode. Click "OK" and move the cursor to the "Preview" box. Clicking on the "Partial" box and then the "Preview" button will give you a preview of how the outside borders of the drawing will be oriented on the paper. Using the "Full" box and the "Preview" button will show you a full detailed drawing and its orientation on the paper. If you are satisfied with the way the drawing will print on the paper click on the "OK" button in the dialog box. The dialog box will close and the command line will direct you to position the paper in the printer or plotter. When ready to print press the "Enter" key and the drawing will be sent to the printer. The printed copy of the drawing will not have the dots from the grid on the paper.

SUMMARY

In summary, one must know or do the following.

1. Know the elements of the graphic window.

2. Understand the command line options.

3. Issue commands to the program.

4. Effectively use the toolbox buttons and drop-down menus.

5. Create layers for different elements of a drawing.

6. Load linetypes into a drawing.

7. Save drawings to a file.

8. Plot or print a drawing.

EXERCISES

EXERCISE 1-1

Name: _____

Copy the two-view sketch. Copy the lamp at 1½ times its given size.

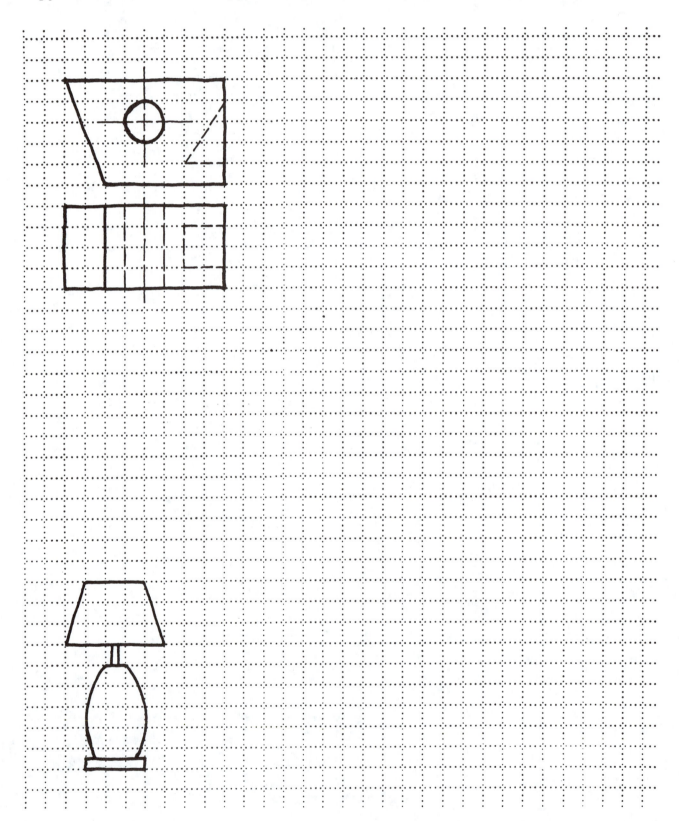

Name: _____

Copy the drawing at 1½ times its size.

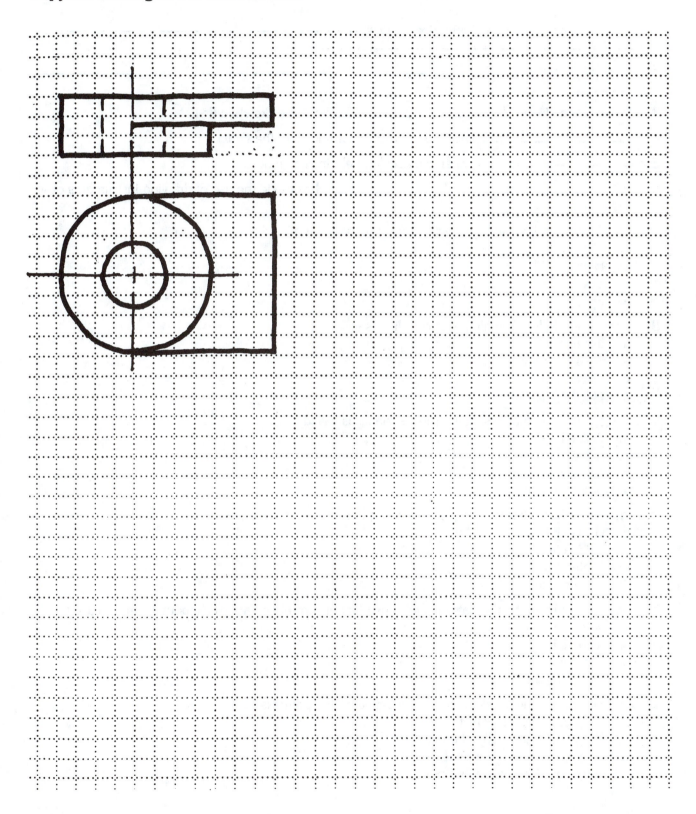

EXERCISE 1-3

Name: _____

Fill in the blanks or circle the correct answer.

1. It is easier to make the sketch if the paper is _____ _____
   _____ _____.

2. Sketches are used to _____ , _____ , and _____
   _____ ideas.

3. In sketching circles, the _____ lines are sketched first.

4. _____ lines are the most prominent on the sketch.

5. _____ paper is a decided help in sketching.

6. _____ _____ has made the use of drawing instruments obsolete.

7. Which of the following lead grades is used for sketching?
   a. 3H.
   b. HB.
   c. 5B.
   d. FH.

8. The basic rule for sketching a straight line is to:
   a. use a ruler.
   b. push the pencil.
   c. press hard.
   d. pull the pencil.

9. The type of line that describes visible surfaces is called a:
   a. visible line.
   b. hidden line.
   c. centerline.
   d. phantom line.

10. The type of line that locates the center of a circle or arc is called a:
    a. visible line.
    b. hidden line.
    c. centerline.
    d. phantom line.

11. The type of line that alternates a short dash (about 3 mm) and a space (about 1 mm) is called a:
    a. visible line.
    b. hidden line.
    c. centerline.
    d. phantom line.

12. The best technique for sketching a line between two points is to:
    a. keep your eye on your thumb.
    b. keep your eye on the starting point.
    c. keep your eye on the pencil point.
    d. keep your eye on the end point.

250

13. On 8½ by 11 inch paper, a tiny gear in a watch would most likely be sketched:
  a. half size.
  b. full size.
  c. double size.

14. On 8½ by 11 inch paper, a 10-inch frying pan would most likely be sketched:
  a. half size.
  b. full size.
  c. double size.

15. On 8½ by 11 inch paper, a typical hand-held calculator would most likely be sketched:
  a. half size.
  b. full size.
  c. double size.

EXERCISE 1-4

Name: _____

Sketch a side view of your desk chair.

EXERCISE 2-1

Name: _____

Fill in the blanks.

1. The space between letters in a word should be the same _____ for all letters.

2. The vertical space between lines of lettering is the same as the height of the _____.

3. The _____ is above center in the letter A.

4. The letter _____ would be a good example to use in illustrating stability of letters.

5. Numerals are made the same height as _____ .

6. Guidelines are made with an _____ grade lead.

7. Strokes are made from _____ _____ _____ and _____ _____ _____ for a right-handed person.

8. Stability is an important factor in the numerals _____ and _____ .

9. The usual height of letters is _____ mm.

10. An _____ -grade lead is used for lettering.

11. Letters and numerals fall into two groups: _____ _____ and _____ _____ .

12. The widest letter of the alphabet is _____ .

13. The space between words in a sentence is equal to the letter _____ .

EXERCISE 2-2

Name: ____________________

Repeat each letter and numeral thrice. Copy each sentence twice.

A	I	N	X
E	K	T	Y
F	L	V	Z
H	M	W	B
C	O	S	3
D	P	U	4
G	Q	0	5
J	R	2	6
7	8	9	

A PERSON CAN LEARN TO LETTER

TOP TO BOTTOM LEFT TO RIGHT

Copy each sentence twice.

THE METRIC SYSTEM IS USED

FOLLOW THE ORDER OF STROKES

PULL THE PENCIL, DON'T PUSH

STABILITY IS AN IMPORTANT FACTOR

THE GRID SERVES AS GUIDELINES

$\frac{1}{2}$ INCH IS APPROXIMATELY 14 MM

Sketch the three orthographic views of the objects. Label folding lines. Use grid to obtain proportion.

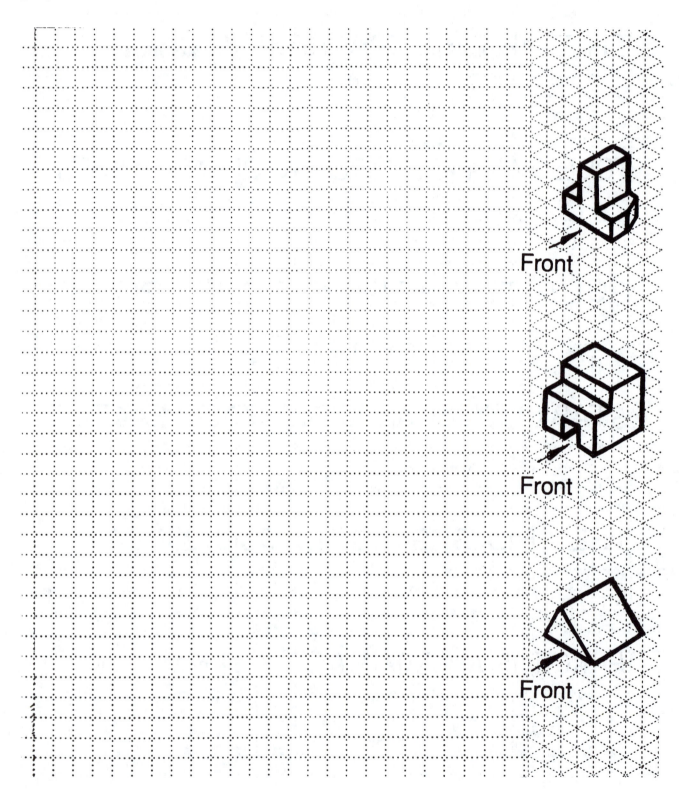

Front

Front

Front

Sketch the three orthographic views of the objects. Label folding lines. Use grid to obtain proportion.

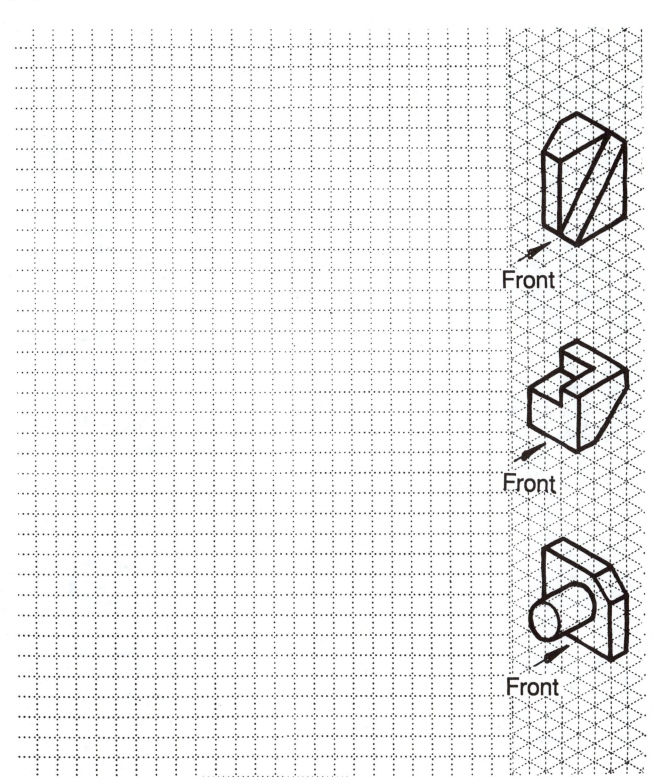

Front

Front

Front

Sketch the three orthographic views of the objects. Label folding lines. Use grid to obtain proportion. The holes go through the objects.

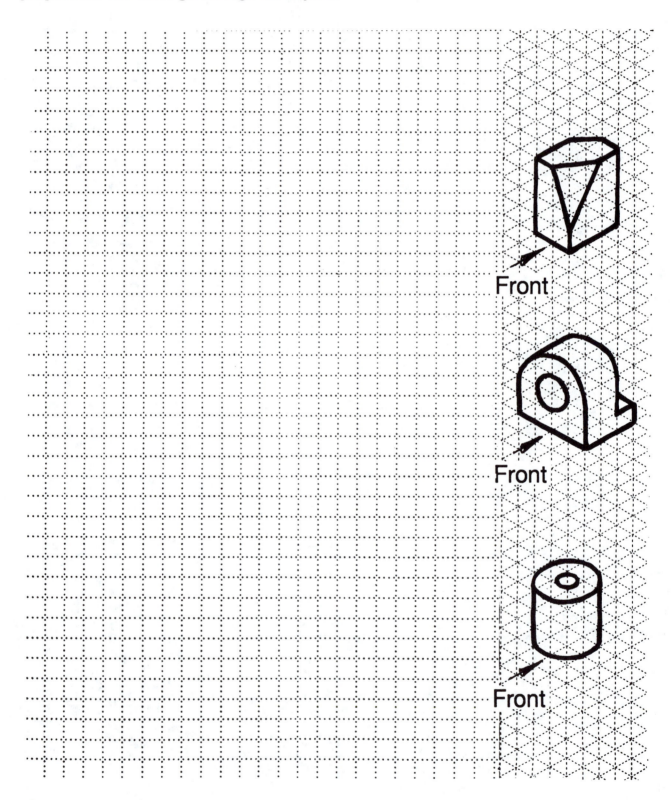

Front

Front

Front

EXERCISE 3-4

Name: _____

Sketch the three orthographic views of the objects. Label folding lines. Use grid to obtain proportion. The holes go halfway through the objects and are made by a drill.

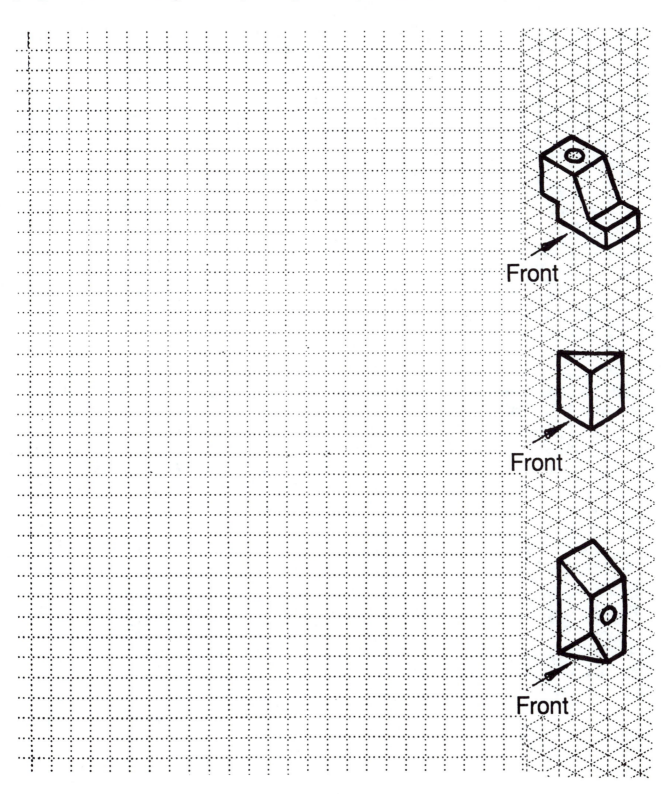

Front

Front

Front

Name: _____

Sketch the three-views of the normal lines of Figure 3-22a and b. The line is 25 mm long. Each square is 5 mm.

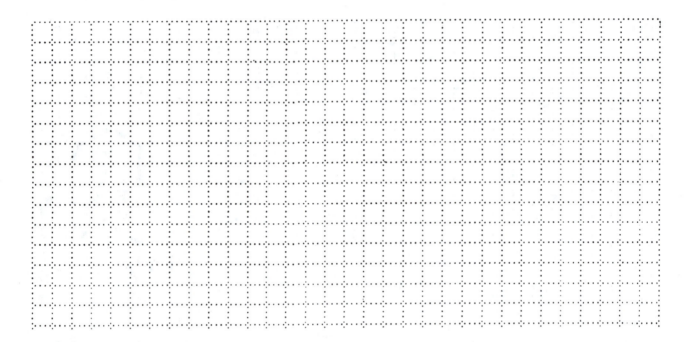

Sketch the three views of the inclined lines of Figure 3-25a and b. The line is 20 mm long and is at a 45° angle.

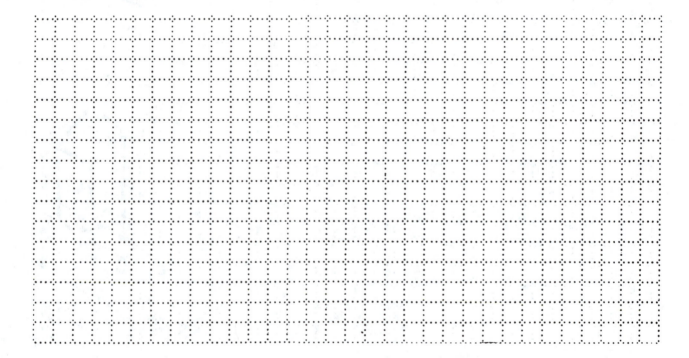

Given two views of a normal surface, sketch the third. Use the 45° line where appropriate. Number all points. Label folding lines. The folding lines cross at the +.

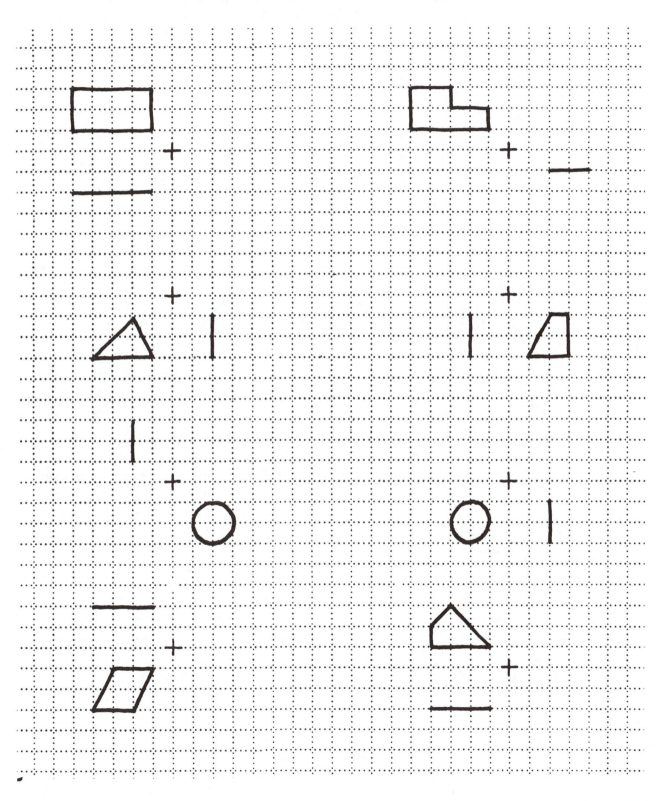

Given two views of an inclined or an oblique surface, sketch the third. Use the 45° line where appropriate. Number all points. Label folding lines. The folding lines cross at the +.

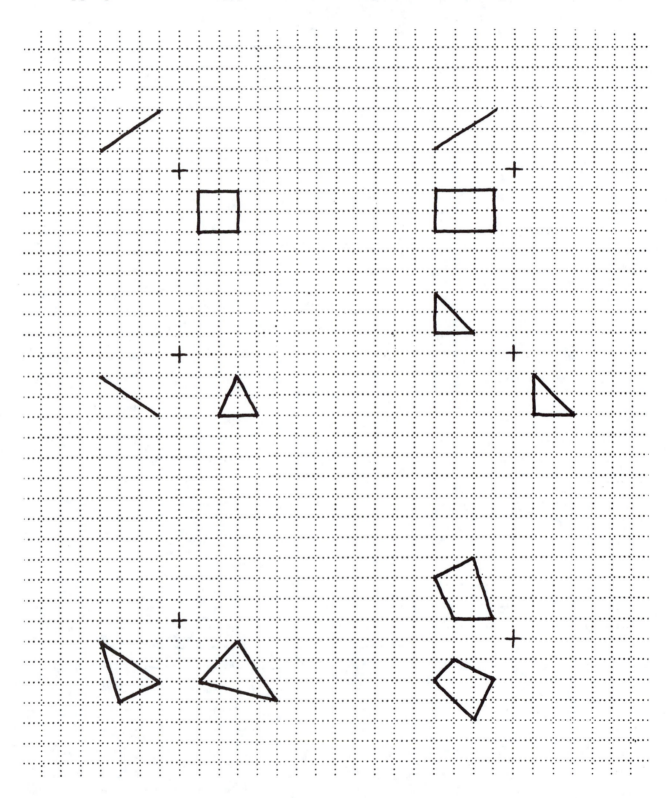

EXERCISE 3-8

Name: _____

Identify the surfaces: normal, inclined, oblique, and cylindrical. Label the surface in the other views as illustrated by the surface labeled "A."

Surface identification

A_____

B_____

C_____

D_____

Surface identification

A_____

B_____

C_____

D_____

EXERCISE 3-9

Name: _____

Identify the surfaces: normal, inclined, oblique, and cylindrical. Label the surface in the other views as illustrated by the surface labeled "A."

Surface identification

A _____

B _____

C _____

D _____

Surface identification

A _____

B _____

C _____

D _____

Given two views, sketch the third. Label folding lines. The folding lines cross at the +.

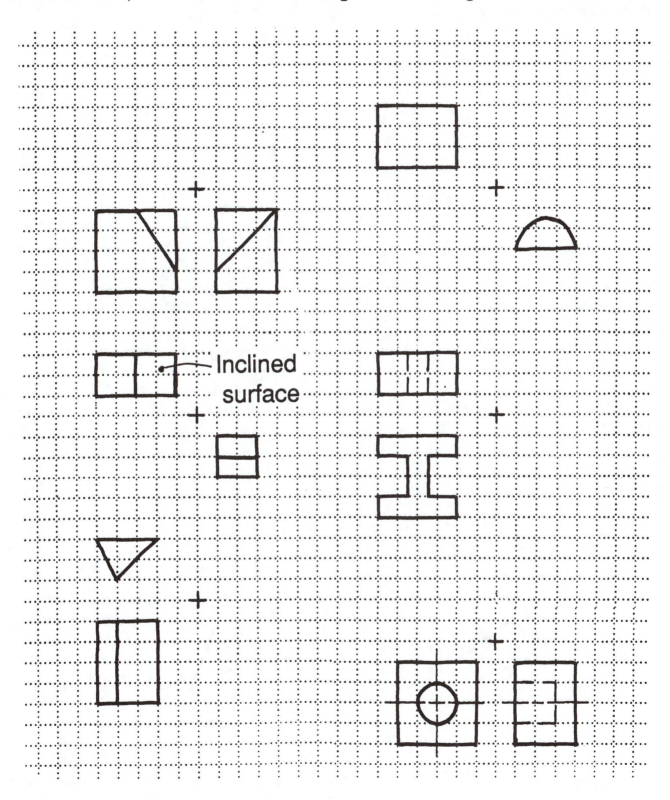

Inclined surface

Name: _____

Given two views, sketch the third. Label folding lines. The folding lines cross at the +. The views given are complete.

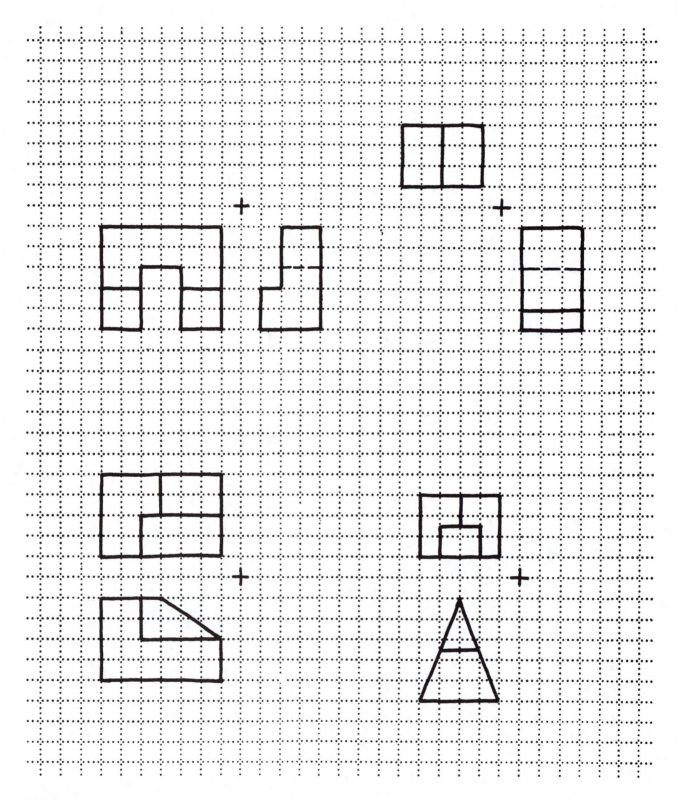

EXERCISE 3-12

Name: _____

Given two views, sketch the third. Label folding lines. The folding lines cross at the +. The views given are complete.

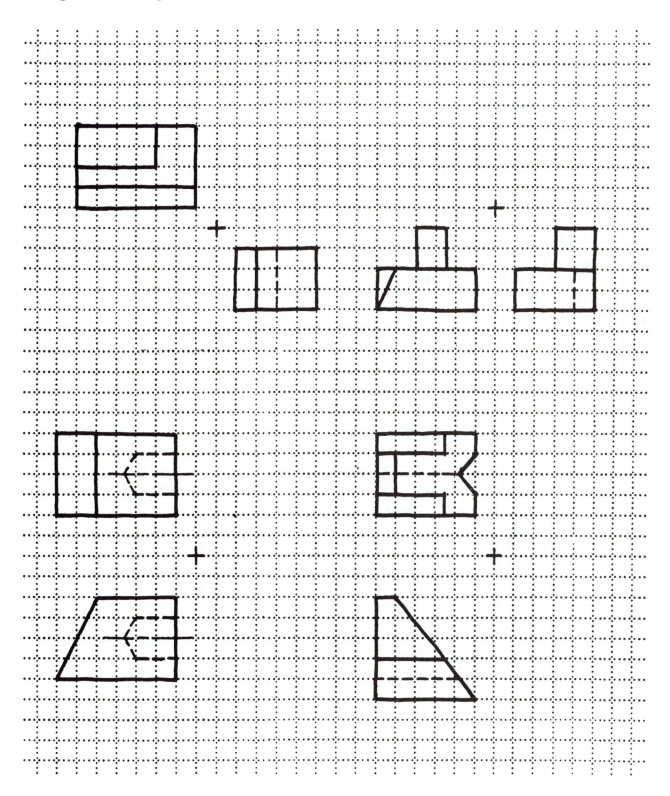

Given two views, sketch the third. The views given are complete.

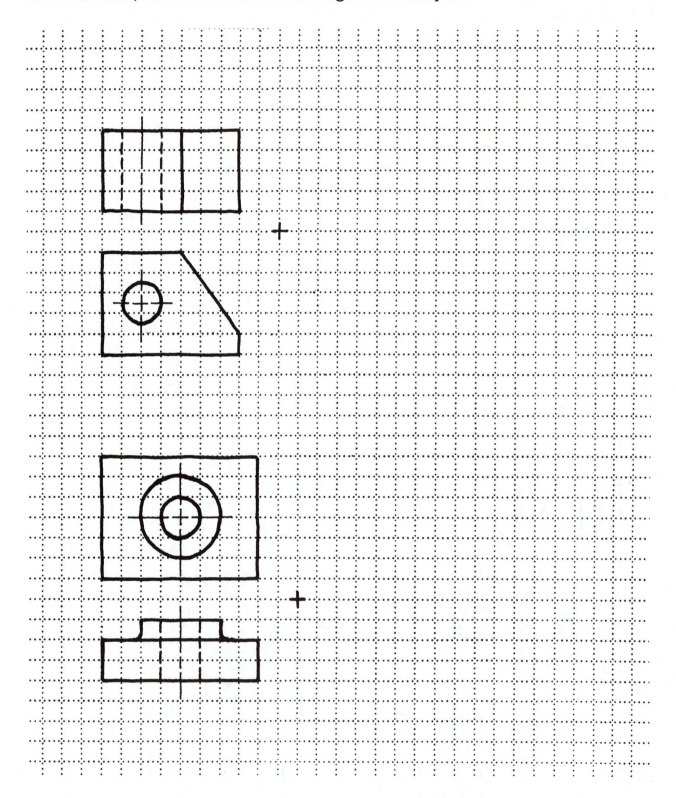

Given two views, sketch the third. The views given are complete.

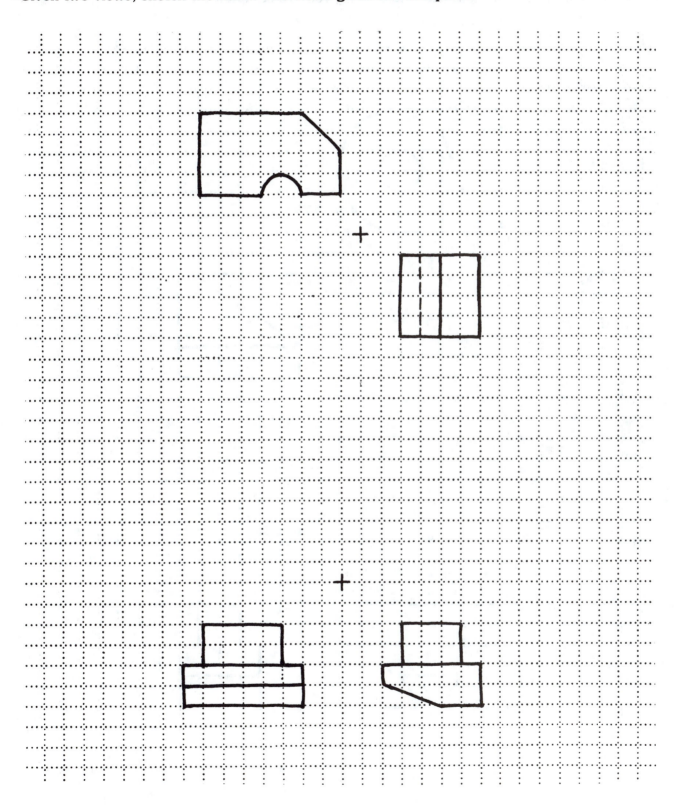

Given two views, sketch the third. The views given are complete.

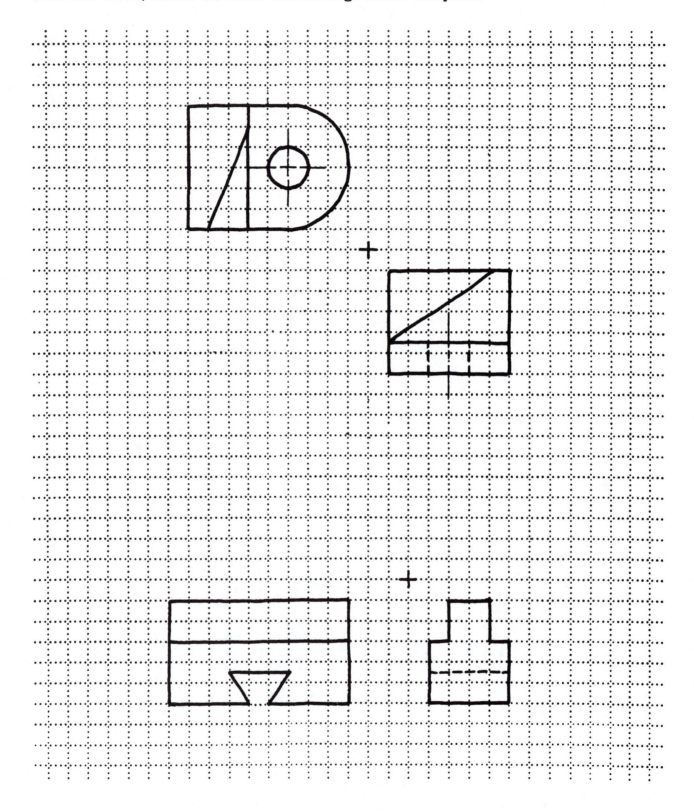

Given two views, sketch the third. The views given are complete.

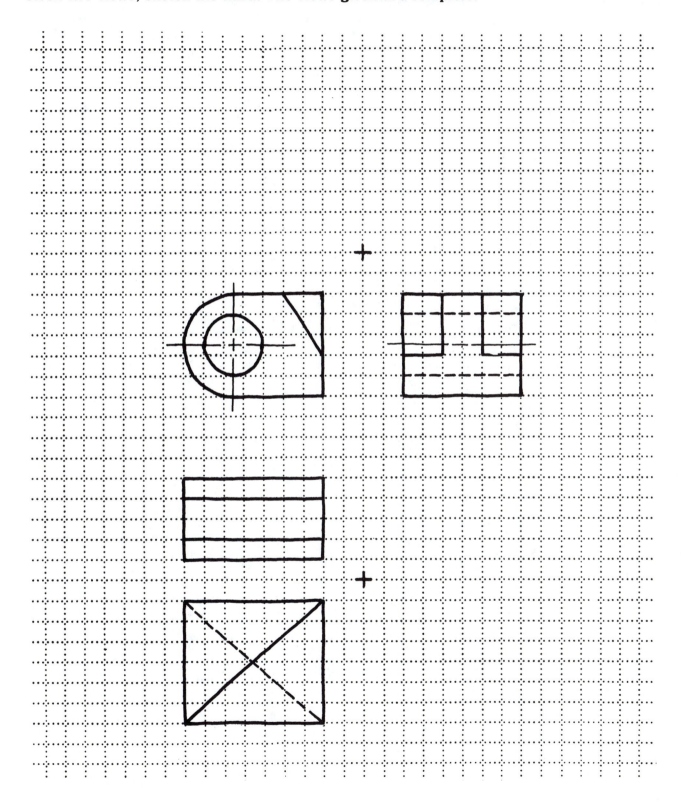

EXERCISE 3-17

Name: _____

Given two views, sketch the third. The views given are complete.

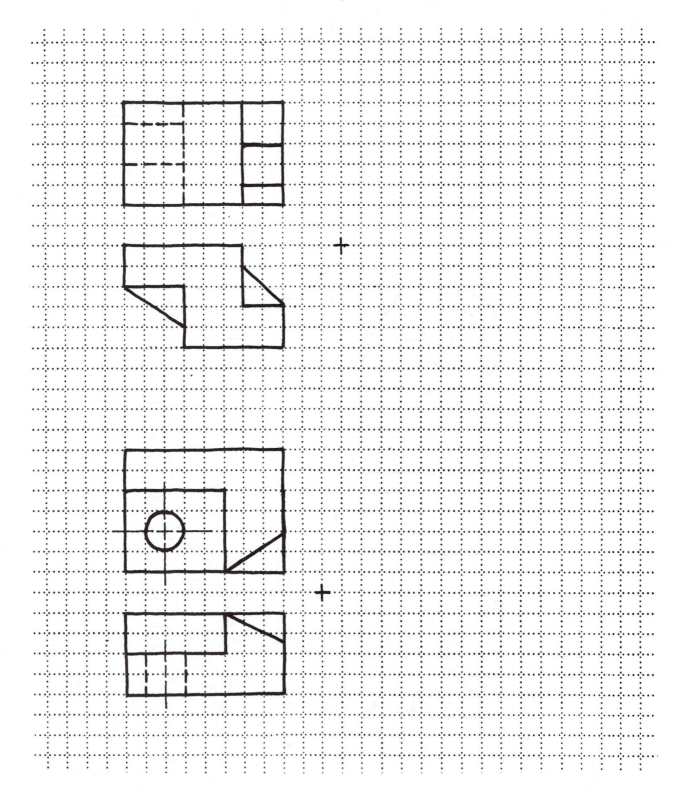

EXERCISE 3-18

Given two views, sketch the third. The views given are complete.

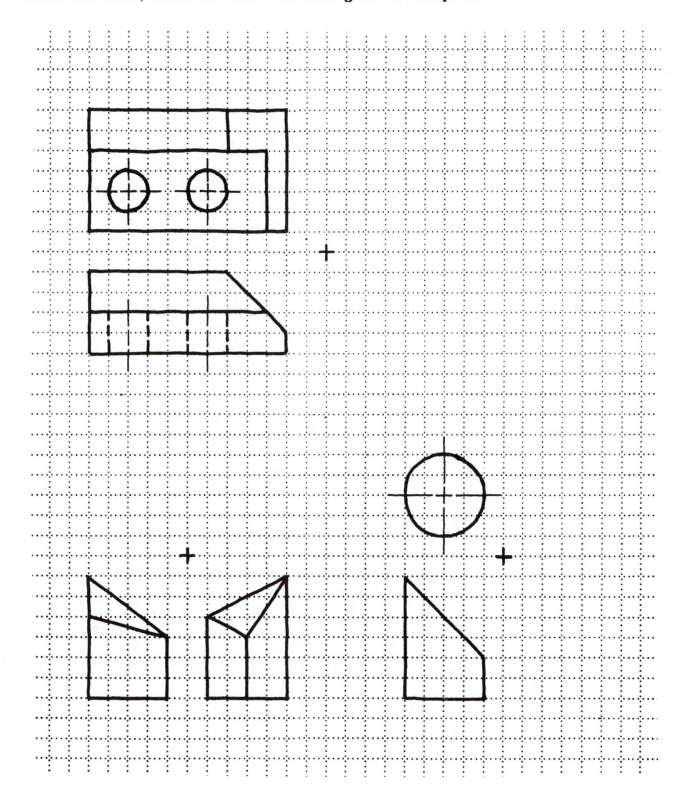

Given two views, sketch the third. The views given are complete.

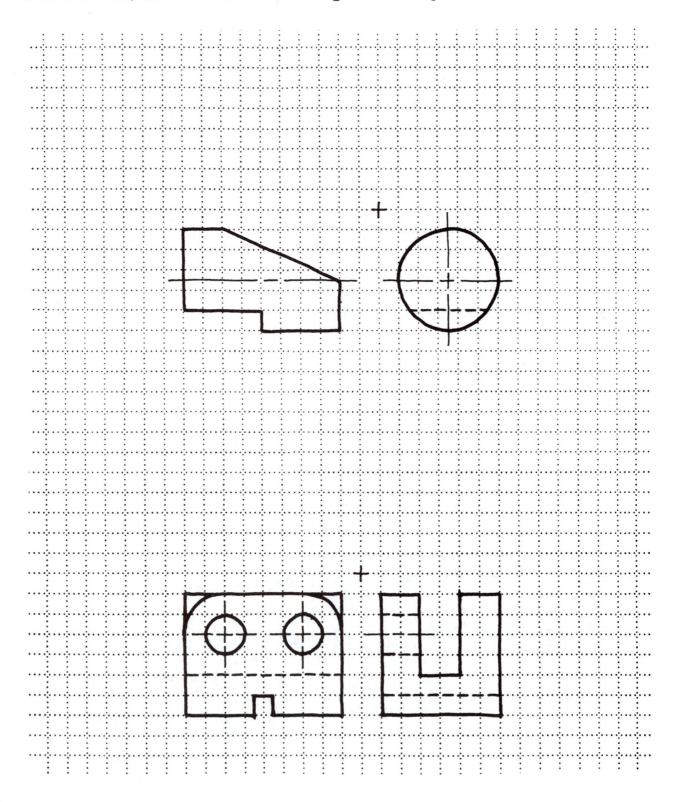

Given two views, sketch the third. The views given are complete.

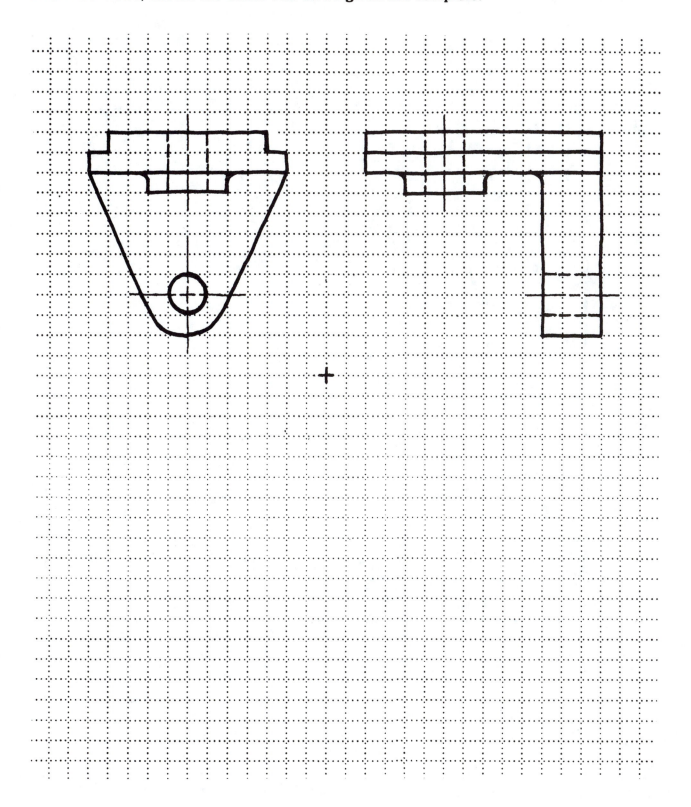

Given two views, sketch the third. The views given are complete.

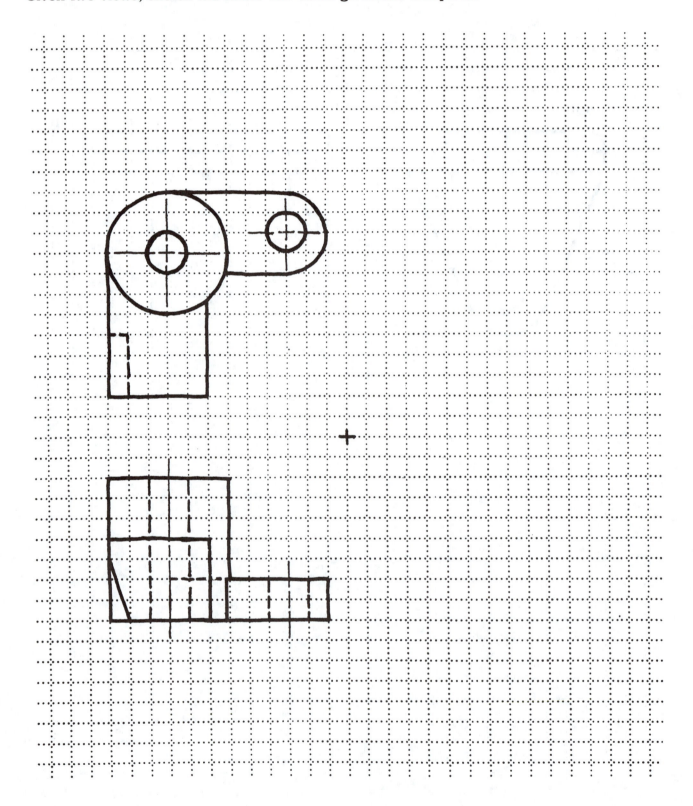

EXERCISE 3-22

Name: _____

Fill in the blanks or circle the correct answer.

1. In orthographic projection, the height of an object is shown in the _____ and _____ views.

2. In orthographic projection the _____ is at an infinite distance.

3. _____ related views are usually necessary to describe an object.

4. Hidden lines take precedence over _____ lines.

5. _____ projection is the basis of multiview sketches.

6. In orthographic projection, the _____ and _____ views are placed horizontal to each other.

7. Looking perpendicularly at the front plane, the _____ and _____ projection planes appear as edges.

8. The _____ view is a view looking down on the object.

9. A projector is _____ to the projection plane.

10. The _____ view is the principal view.

11. In orthographic projection, the _____ of an object is shown in the top and side views.

12. In orthographic projection, the width of an object is seen in the _____ and _____ views.

13. Orthographic projection is _____ _____ projection.

14. A hidden line takes precedence over a _____ .

15. The angle, expressed in degrees, between adjacent projection planes is _____ .

16. The view showing the most descriptive shape of the object should be selected as the _____ view.

17. A numbering system can be helpful in projecting _____ .

18. Orthographic projection requires that the projectors be:
    a. parallel to the plane of projection.
    b. perpendicular to the plane of projection.
    c. oblique to the plane of projection.
    d. None of the above.

19. Each view in orthographic projection shows only:
    a. one direction/dimension.
    b. two directions/dimensions.
    c. three directions/dimensions.
    d. four directions/dimensions.

20. The folding line between the top and front projection plane is labeled:
    a. T/F.
    b. F/S.
    c. T/S.

21. Which of the following lines is the most prominent on a sketch?
    a. Center.
    b. Right.
    c. Hidden.
    d. Visible.

22. The type of plane surface that is parallel to one of the projection planes is called:
    a. normal.
    b. oblique.
    c. parallel.
    d. inclined.

23. The type of plane surface that is perpendicular to one of the projection planes and inclined to the others is called:
    a. normal.
    b. oblique.
    c. parallel.
    d. inclined.

24. The type of plane surface that is neither parallel nor perpendicular to a projection plane is called:
    a. normal.
    b. oblique.
    c. parallel.
    d. inclined.

25. When viewing an object from the front, the side and top projection planes appear as a/an:
    a. perpendicular.
    b. cylinder.
    c. edge.
    d. surface.

26. A line that is parallel to two projection planes is called:
    a. normal.
    b. oblique.
    c. parallel.
    d. inclined.

27. A line that is parallel to one of the projection planes and inclined to the others is called:
    a. normal.
    b. oblique.
    c. parallel.
    d. inclined.

28. A line that is neither parallel nor perpendicular to any of the projection planes is called:
   a. normal.
   b. oblique.
   c. parallel.
   d. inclined.

Given the orthographic views of the two objects, sketch the isometric.

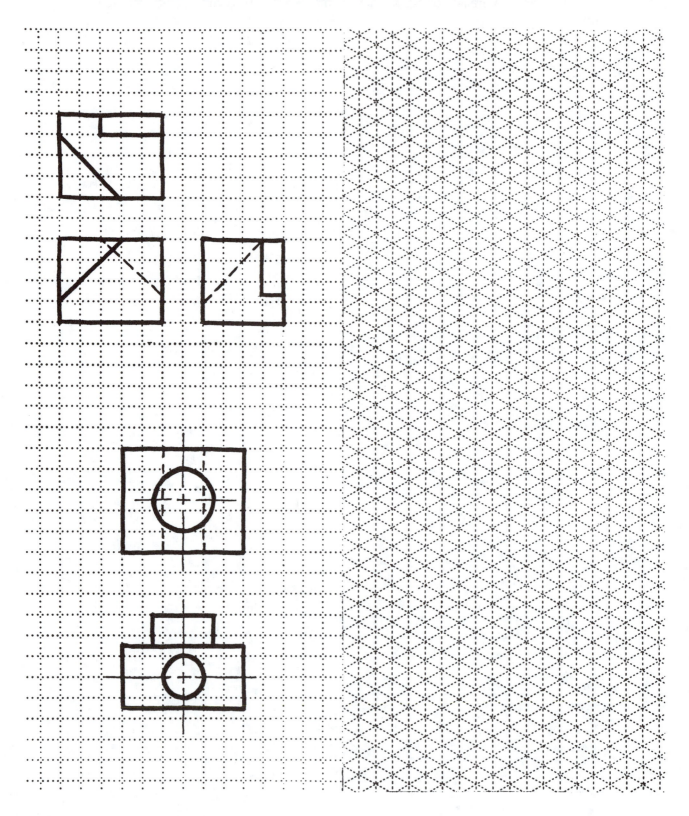

Given the orthographic views of the two objects, sketch the isometric.

Name: _____

Given the orthographic views of the two objects, sketch the isometric.

EXERCISE 4-4

Name: _____

Construct an isometric sketch of a common object that has at least one inclined and one circular surface.

Given the orthographic views of the two objects, sketch the cabinet oblique.

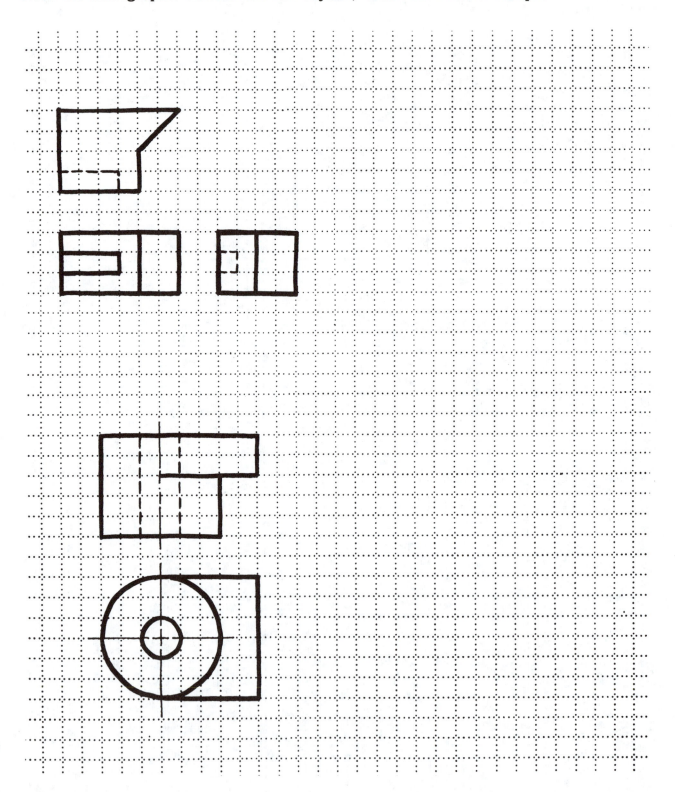

Name: _____

Given the orthographic views of the two objects, sketch the cabinet oblique.

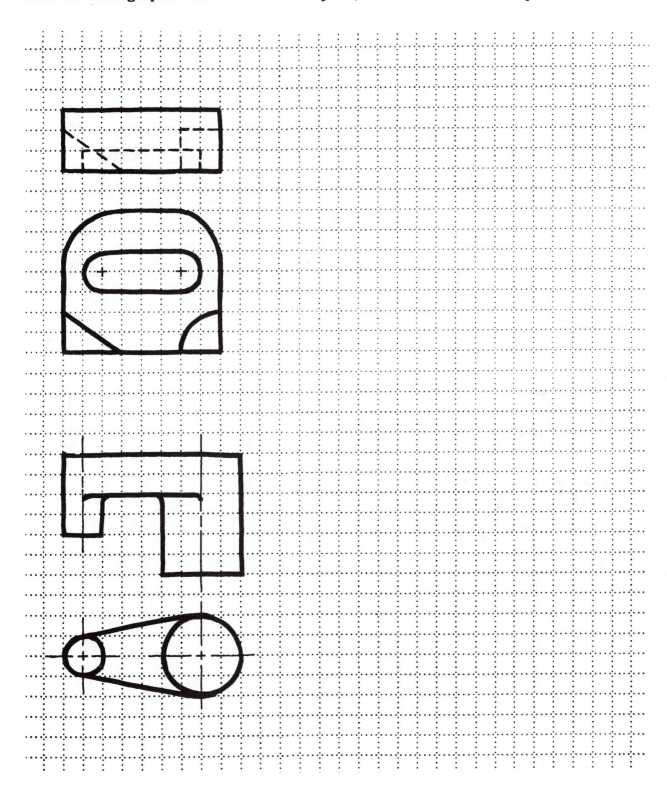

Given the orthographic views of the object, sketch the cabinet oblique.

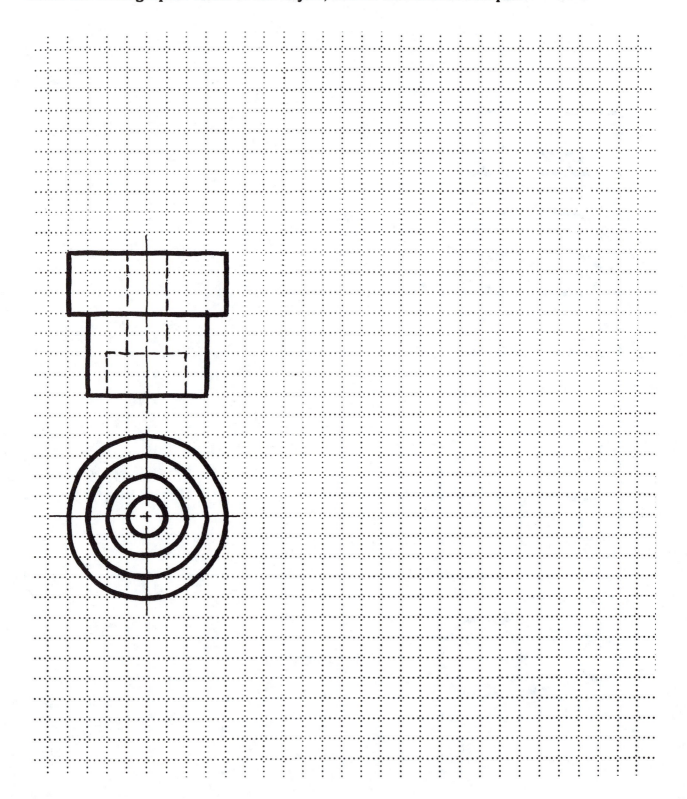

EXERCISE 4-8

Name: _____

Construct a cabinet oblique sketch of a common object that has an irregular or circular surface.

EXERCISE 4-9

Name: _____

Fill in the blanks or circle the correct answer.

1. In axonometric projection the object is in a _____ position.

2. Isometric projection uses _____ projection planes.

3. The observer in orthographic and axonometric projection is at a/an _____ distance.

4. To construct circles in the isometric, the _____ method is recommended.

5. In cabinet oblique sketches, the length along the 45° receding axis is reduced _____.

6. Circles will appear as true circles on _____ surfaces of an oblique sketch.

7. In isometric, irregular shapes require the plotting of a series of points with each point determined by _____ measurements.

8. A true isometric projection is foreshortened by _____ .

9. In axonometric projection, the parallel lines of sight are _____ to the projection plane.

10. The most important sketching technique for an isometric is the _____ grid.

11. Oblique sketching has the advantage that the front surface and all parallel surfaces are sketched _____ _____ as the front orthographic view.

12. The position of the object in isometric should be accomplished to maximize the number of _____ lines.

13. An orthographic square has two angles at 120° and two at _____ .

14. In pictorial sketches, _____ lines are usually omitted.

15. Circles will appear as _____ on surfaces in isometric.

16. _____ oblique sketches are recommended.

17. _____ or _____ surfaces of the object should be selected for the front face of an oblique sketch.

18. For isometric projection the object is _____ and _____ .

19. _____ may not be measured by their degrees in isometric sketches.

20 A reversed isometric axis occurs when we are looking _____ at an object.

21. In an oblique sketch _____ axes are at 90° to each other.

22. _____ projection is the most widely used because the axes angles and lengths are equal.

23. In a cabinet oblique sketch:
    a. two axes are placed at 90° to each other, and the receding axis at 45° to one of the 90° axes.
    b. all axes are equally spaced at 120°.

   c. two axes are placed at 90° to each other, and the receding axis at 135° to one of the 90° axes.

   d. two axes are placed at 90° to each other, and the receding axis at 135° to one of the 90° axes and 45° to the other 90° axis.

24. Which of the following graphic presentations *cannot* be understood by the layman?
   a. Orthographic.
   b. Axonometric.
   c. Oblique.
   d. Perspective.

25. Vertical measurements in isometric can only be made on or parallel to the:
   a. vertical axis.
   b. inclined axis.
   c. horizontal axis.
   d. oblique axis.

26. In a cavalier oblique sketch:
   a. the receding axis is drawn at 45° and the length reduced one-half.
   b. the receding axis is drawn at 0° to 90° and the length on the receding axis reduced one-half.
   c. the receding axis is drawn at 45° and the length is full size.
   d. the receding axis is drawn at 0° to 90° and the length on the receding axis reduced one-half.

27. The relationship of the axes to each other in an isometric sketch is:
   a. 30°.
   b. 60°.
   c. 90°.
   d. 120°.

28. In a general oblique sketch, the receding axis is:
   a. perpendicular to one of the principal axes.
   b. parallel to one of the principal axes.
   c. at an angle of 0° to 90° with one of the principal axes.
   d. None of the above.

29. The type of axonometric projection that has two axes angles equal is called:
   a. dimetric.
   b. trimetric.
   c. isometric.
   d. cabinet.

30 One of the axes in an isometric sketch is *usually:*
   a. horizontal.
   b. parallel.
   c. vertical.
   d. None of the above.

Sketch full-sectional views of the objects. Sketch the front view as a full section. Be sure to label.

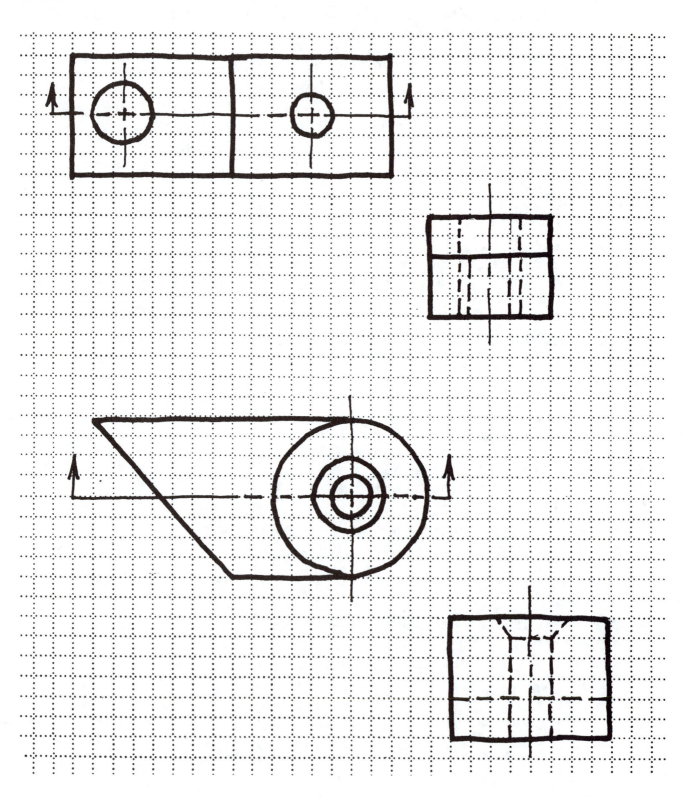

Sketch full-sectional views of the objects. Label the sectional view.

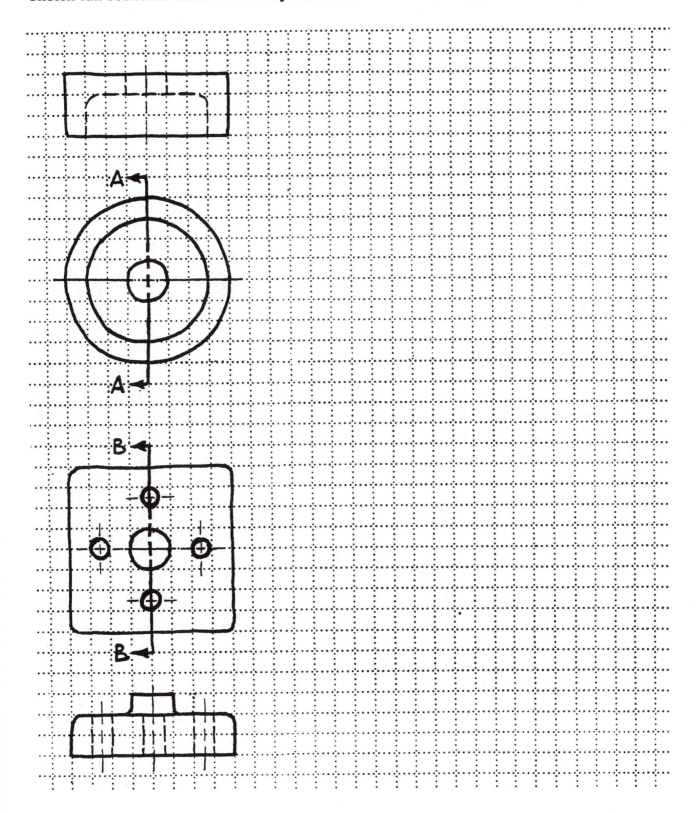

EXERCISE 5-3

Name: _____

Sketch full-sectional views of the objects. Label the sectional view.

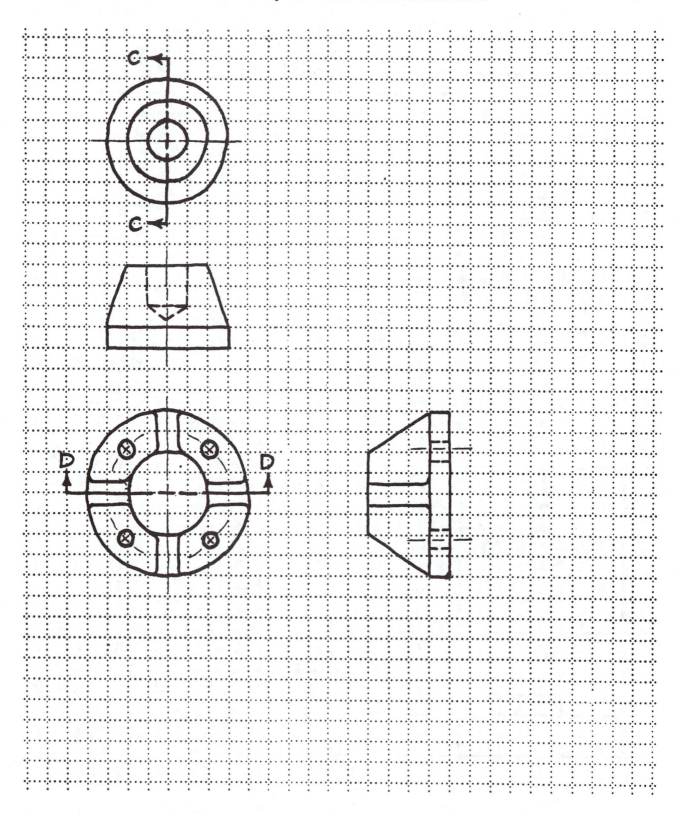

Name: _____

Sketch full-sectional views of the objects. Label the sectional view.

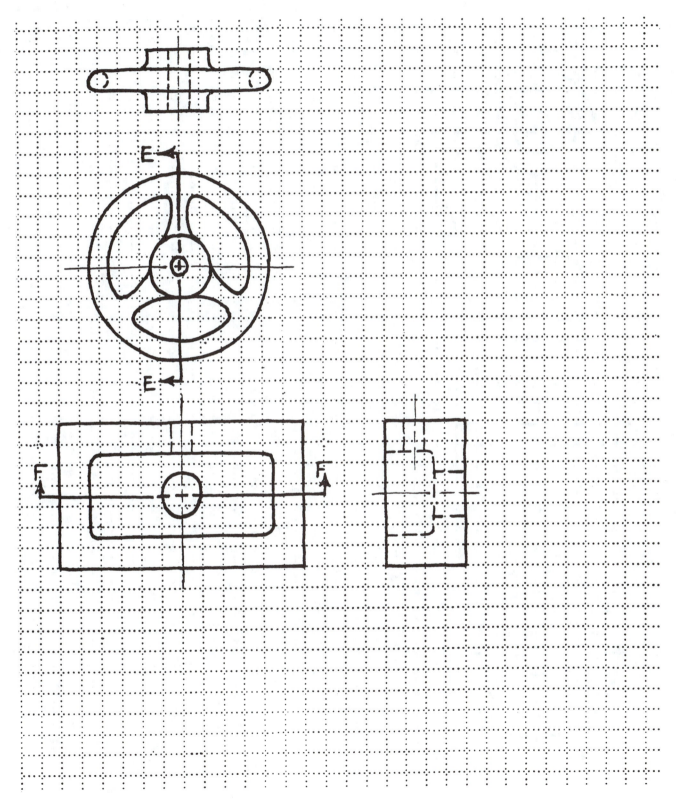

EXERCISE 5-5

Name: _____

Sketch half-sectional views of the objects. Label the sectional view.

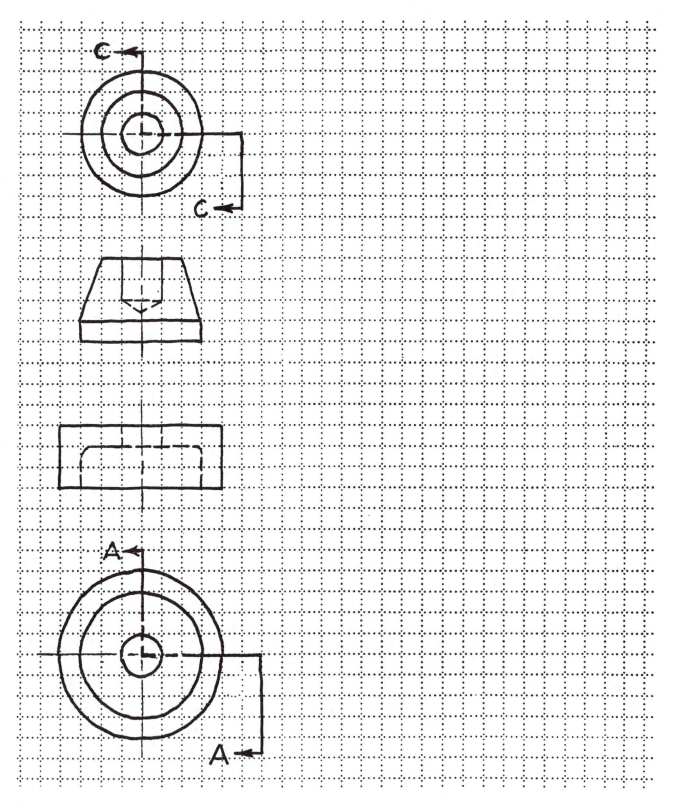

Name: _____

Sketch half-sectional views of the objects. Label the sectional view.

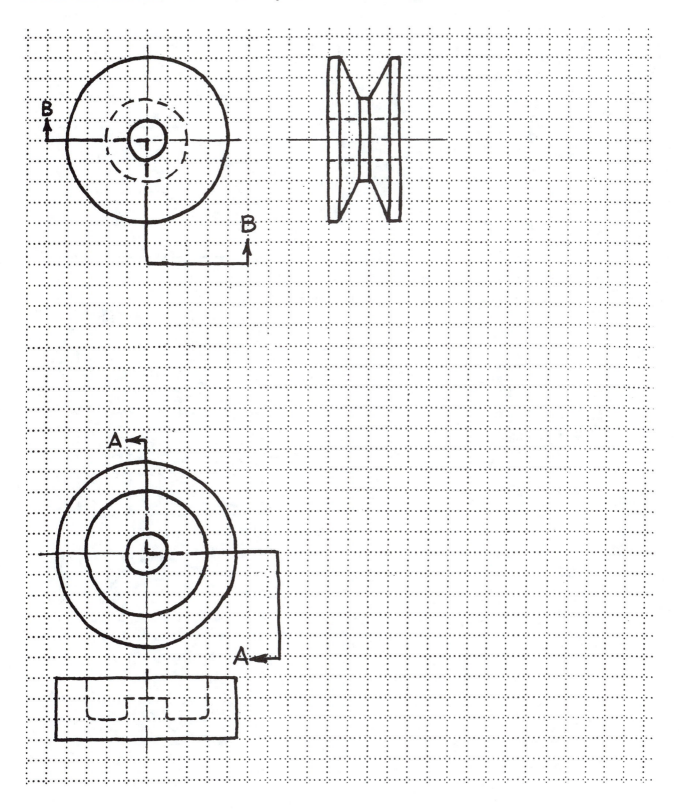

Name: _____

Identify the different types of sectional views.

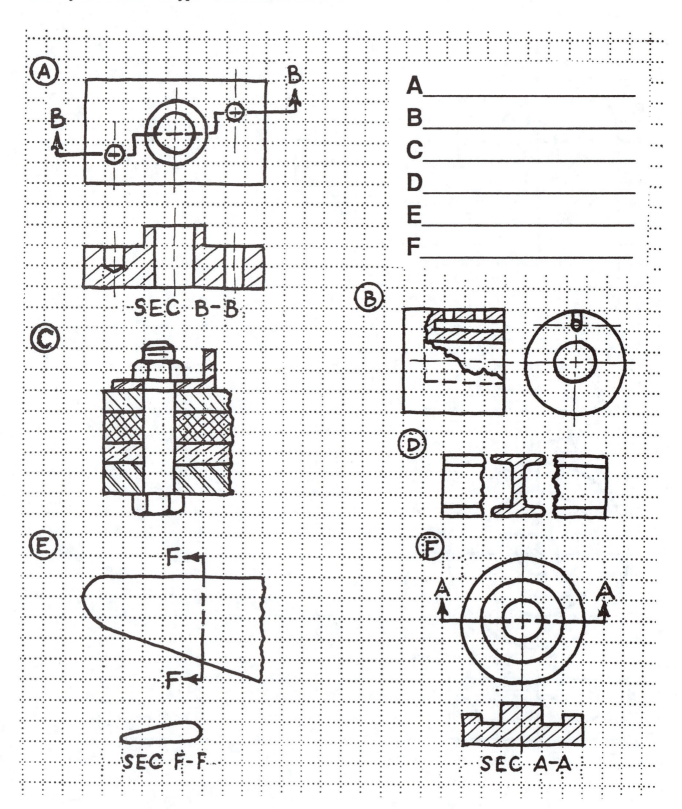

A _____

B _____

C _____

D _____

E _____

F _____

SEC B-B

SEC F-F

SEC A-A

Fill in the blanks.

1. It is considered _____ practice to show hidden lines in the sectional view.

2. The cutting plane for a half-section removes _____ of the object.

3. Detail or removed sections may be drawn to an _____ scale in order to emphasize detail.

4. The angle most commonly used for drawing crosshatched lines is _____ °.

5. "Conventional practice" is not a _____ projection. It is used because it makes the sketch easier to understand.

6. When the cutting plane passes through and parallel to the axis of a bolt or shaft, it should not be _____ .

7. In half-section views, _____ lines are generally not included on unsectioned parts.

8. A _____ section is used as part of an exterior view to show some detail without sketching a complete section.

9. A cutting plane is an _____ plane used to show the path of cutting an object to make a sectional view.

10. A cutting plane line is made the same thickness as a _____ line.

11. In a sectional view, only the _____ in contact with the cutting plane is crosshatched.

12. A _____ or _____ is not crosshatched when the cutting plane passes perpendicular to its axis.

13. The arrows on a cutting plane point in the _____ _____ _____ .

14. The letters on the cutting plane line identify the _____ .

15. A full-sectional view is an _____ view.

16. All _____ lines behind the cutting plane are shown in the sectional view.

17. A _____ _____ line takes precedence over a center line.

18. A _____ section is made directly on an exterior view.

19. The cross section of a bar, spoke, rib, or other shape may be shown directly on the orthographic view by means of a _____ section.

20 Identify the following materials from their symbols.

a _____   b _____

c _____   d _____

e _____   f _____

g _____   h _____

i _____

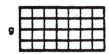

Find the true length of the line. Label all folding lines, points, and true length lines.

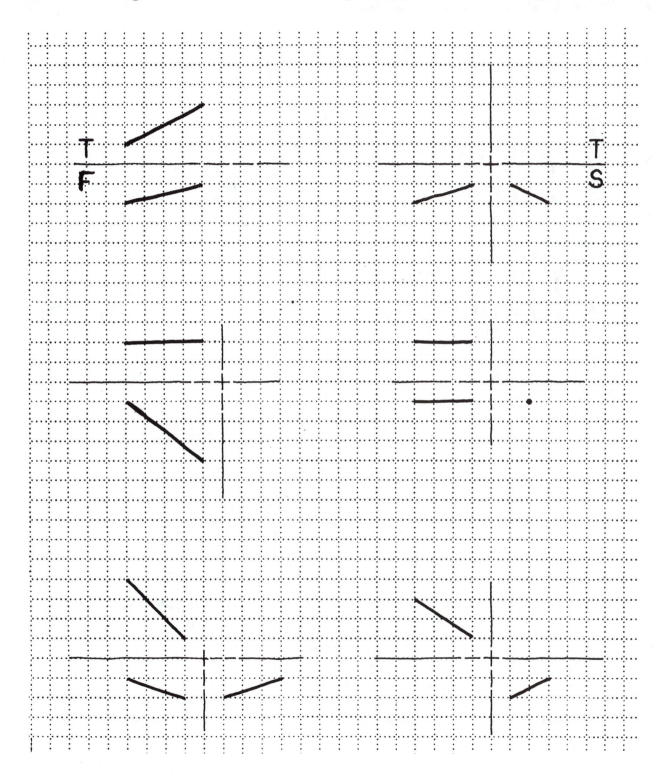

Find the true size of the inclined surface. Label the folding lines and true size surface. For the cylinder, use the axis as the folding line.

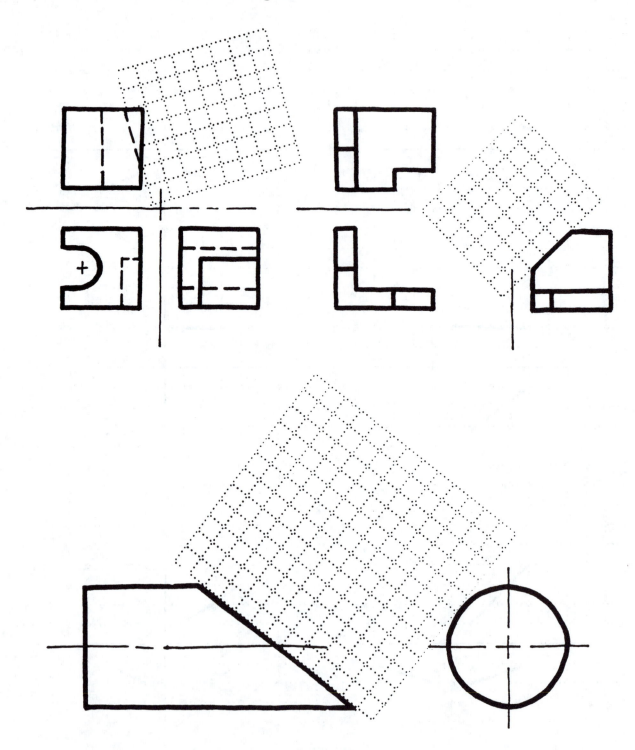

Construct partial auxiliary views for the inclined surfaces. The hole size is 10 mm.

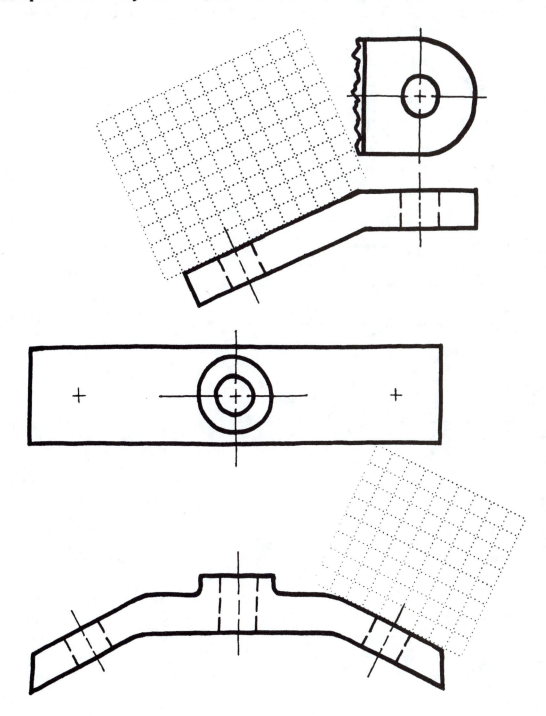

Find the true size of the oblique surface. Take the first auxiliary off the top view in the upper figure and off the side view in the lower one. Use the squares to assist with the projection angle and the distance.

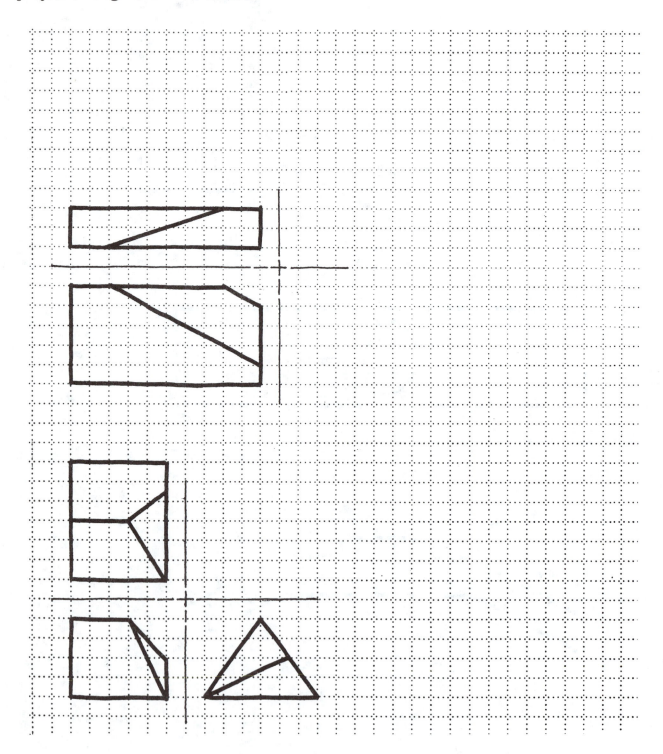

Name: _____

Find the true size of the oblique surface.

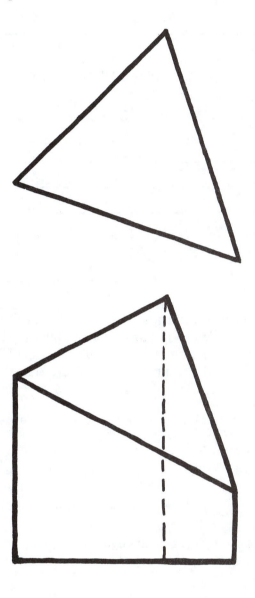

EXERCISE 6-6

Name: _____

Fill in the blanks or circle the correct answer.

1. The three principal _____ _____ are sometimes inadequate to fully describe an object.

2. An _____ number of primary auxiliary views taken off the top view are possible.

3. An auxiliary view may be used to replace a _____ view.

4. The projection of an oblique line to an auxiliary view will show the line in true length provided that the folding line is _____ to the oblique line.

5. It is not necessary to show the complete _____ view.

6. A vertical line will appear as a _____ in the top view.

7. A secondary auxiliary view is necessary to obtain the true size of an _____ surface.

8. Any level (horizontal) line will show _____ _____ in the top view.

9. To obtain the true size of an oblique surface, the secondary auxiliary folding line is placed:
   a. perpendicular to the edge view of the surface in the auxiliary view.
   b. parallel to the edge view of the surface in the auxiliary view.
   c. oblique to the edge view of the surface in the auxiliary view.
   d. None of the above.

10. Auxiliary views are used to find the:
    a. true length of an oblique line.
    b. true size of an inclined surface.
    c. true size of an oblique surface.
    d. All of the above.

11. When looking at the auxiliary projection plane taken off the front projection plane, the:
    a. projection plane appears as an edge.
    b. auxiliary projection plane appears as an edge.
    c. front projection plane appears as an edge.
    d. None of the above.

12. To obtain the true size of an oblique surface, the primary auxiliary folding line is placed:
    a. perpendicular to the top view.
    b. oblique to the true length of a line on that surface.
    c. perpendicular to the true length of a line on that surface.
    d. parallel to the true length of a line on that surface.

13. When a primary auxiliary plane is taken off the front projection plane, it is:
    a. perpendicular to the top projection plane.
    b. parallel to the front projection plane.
    c. perpendicular to the front projection plane.
    d. parallel to the top projection plane.

14. To obtain the true size of an inclined surface, the primary auxiliary projection plane is placed:
    a. parallel to the edge view of that surface.
    b. perpendicular to the edge view of that surface.
    c. inclined to the edge view of that surface.
    d. None of the above.

15. When a primary auxiliary view of an object is taken off the top view, the distance from a point 1 to folding line AT:
    a. is equal to the distance from point 1 in the front view to folding line T/F.
    b. is unequal to the distance from point 1 in the front view to folding line T/F.
    c. is equal to the distance from point 1 in the top view to folding line T/F.
    d. is unequal to the distance from point 1 in the top view to folding line T/F.

16. When looking at the auxiliary projection plane taken off the top projection plane, the:
    a. projection plane appears as an edge.
    b. auxiliary projection plane appears as an edge.
    c. top projection plane appears as an edge.
    d. None of the above.

17. What two types of auxiliary views are normally used?
    a. Principal and secondary.
    b. Alternate and primary.
    c. Principal and primary.
    d. Primary and secondary.

Name: _____

Dimension the objects. Each square is 5 mm. Indicate assumptions you make. All surfaces are finished.

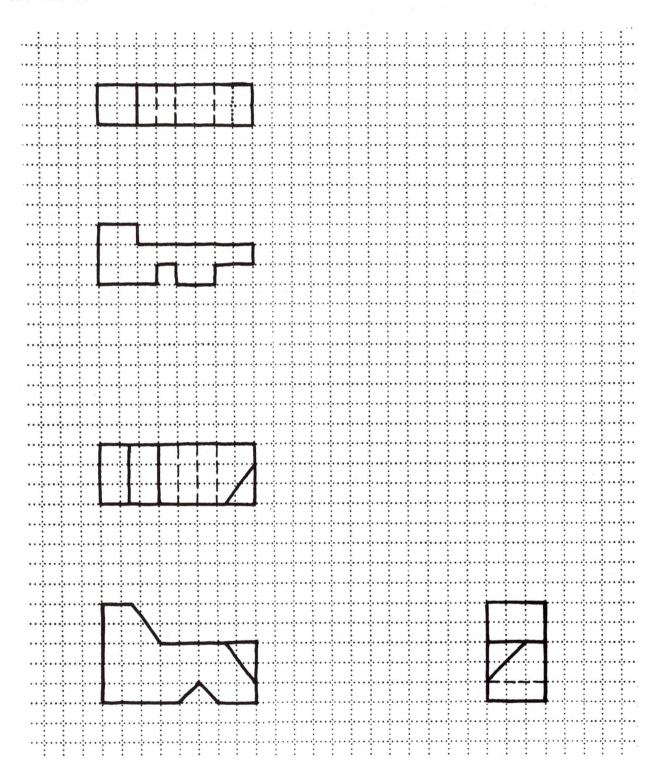

Dimension the objects. Each square is 5 mm. Indicate assumptions you make. All surfaces are finished.

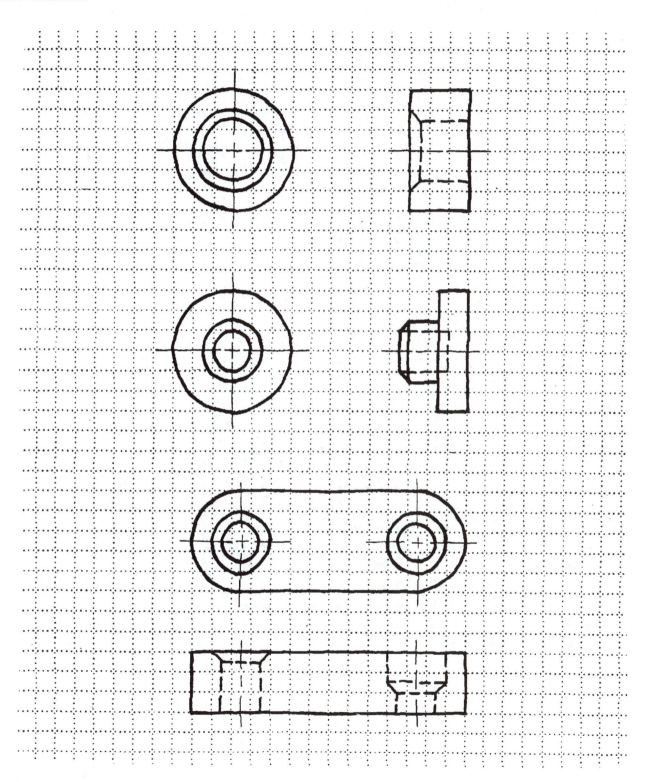

Name: _____

Dimension the object. Each square is 5 mm. Indicate assumptions you make.

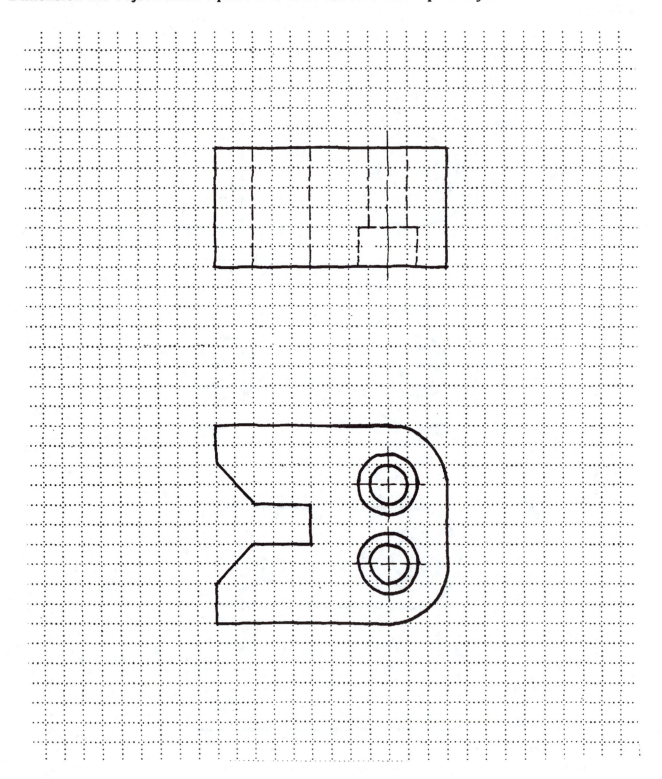

Dimension the object. Each square is 5 mm. Indicate assumptions you make. The object is a casting.

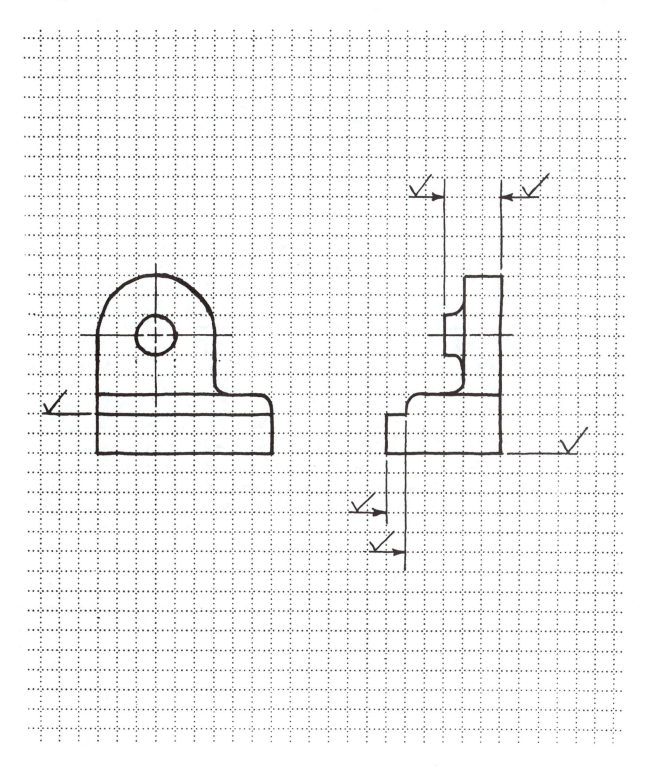

Dimension the objects. Each square is 5 mm. Indicate assumptions you make. All surfaces are finished.

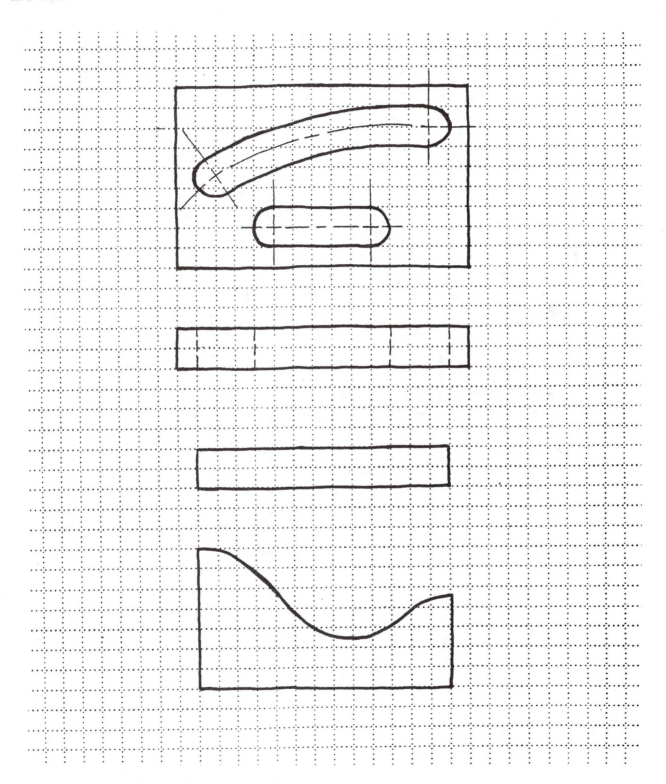

Dimension the objects. Each square is 5 mm. Indicate assumptions you make. The upper figure is a casting and the lower figure is finished all over (FAO).

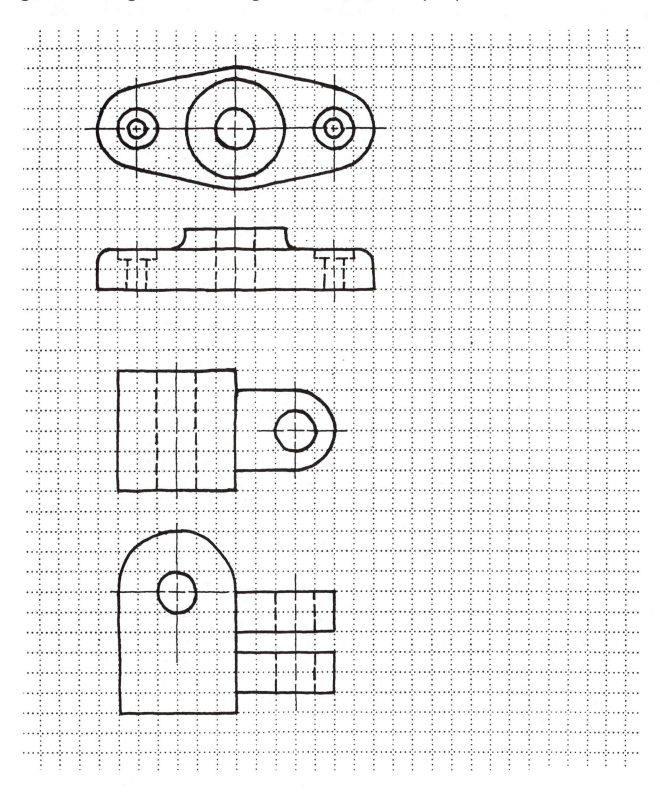

Dimension the objects. Each square is 5 mm. Indicate assumptions you make. Use rectangular coordinates for the upper figure and polar coordinates for the lower. The holes are equally spaced in the lower figure and there are no counterbores of the small holes.

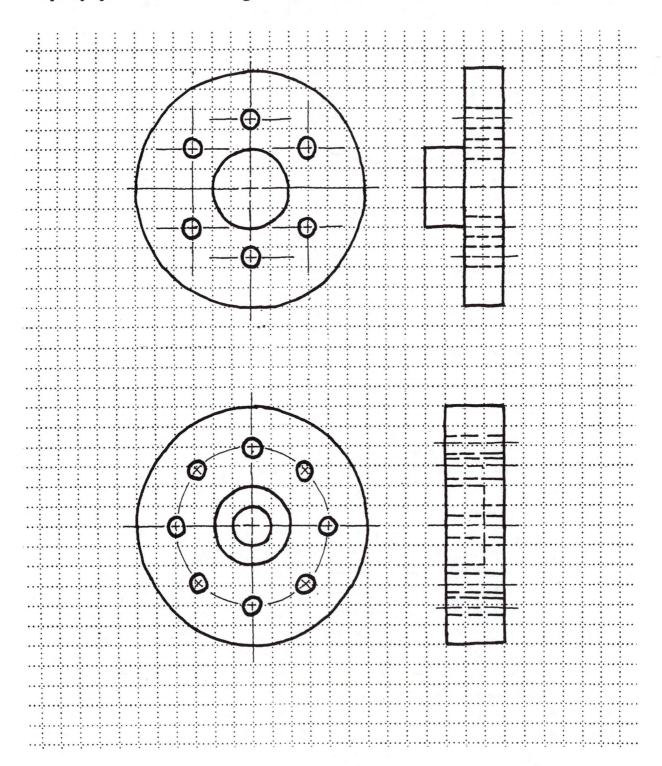

Dimension the objects. Each square is 5 mm. Indicate assumptions you make. For the upper figure, use dimension lines. For the middle figure, omit dimension lines. For the lower figure, use the tabular form.

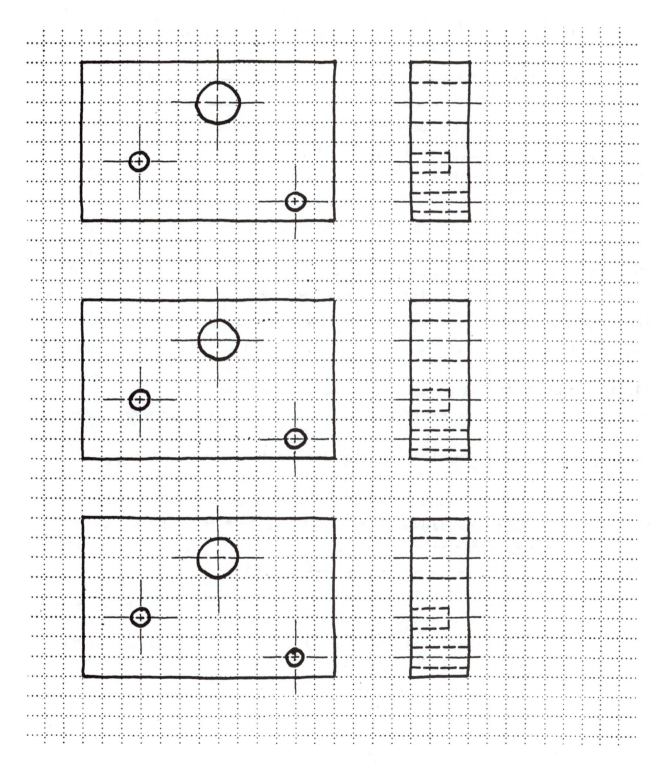

Fill in the blanks or circle the correct answer.

1. A dimension line is a thin line with a break near its center for the _____ .

2. As a general rule, when spacing dimensions from an object, the first dimension line should be _____ from the view, and all succeeding dimensions spaced _____ apart.

3. Special dimensioning techniques are required for _____ spaces.

4. A center line may be extended and used as an _____ line.

5. _____ lines can not be used as extension lines.

6. _____ lines can cross each other.

7. _____ lines should not be located within a view of an object.

8. Dimensioning to hidden lines is considered _____ practice.

9. A _____ is a light continuous line that extends from a note to the feature of a part to which the note applies.

10. A leader is not used with _____ notes.

11. A note that applies to an individual feature on a sketch is called a _____ note.

12. A leader for a hole drill size should terminate at the _____ of the hole.

13. The symbol _____ shown on a drawing indicates that the surface is to be finished.

14. Dimensions should be given to _____ surfaces.

15. In the metric system of dimensioning, a _____ is placed before a value less than one.

16. In general dimensions should be placed outside the view showing the _____ shape of the part.

17. Dimensions are classified as _____ and _____ .

18. The preferred place to specify the diameter of a 12-mm drilled hole is on the view showing the hole as a _____ .

19. It is necessary to group _____ dimensions.

20. The least important dimension in a series should be _____ .

21. A diameter of a cylinder is usually dimensioned in the _____ view.

22. The most accurate method of dimensioning an angle is by _____ at its end points.

23. An arc is dimensioned by its _____ .

24. Normally, an overall dimension is not required with a _____ end part.

25. Countersunk holes are used for _____ _____ screws.

26. If accuracy is critical for the location of holes on a bolt circle, use _____ dimensions.

27. The _____ system of dimensioning is required.

28. Extension and dimension lines are constructed with a:
    a. 2B pencil.
    b. HB pencil.
    c. 3H pencil.
    d. 6H pencil.

29. The proportion, length to width, of an arrowhead is:
    a. 1:1.
    b. 2:1.
    c. 3:1.
    d. 4.1.
    e. 3:2.

30 Dimension lines should be spaced:
    a. 10 mm from the view and 10 mm from each other.
    b. 12 mm from the view and 6 mm from each other.
    c. 12 mm from the view and 12 mm from each other.
    d. 10 mm from the view and 6 mm from each other.

31. In the unidirectional system of dimensioning, dimension figures read from:
    a. the bottom and left side of the sheet.
    b. the bottom only.
    c. the bottom and right side of the sheet.
    d. the right side only.

32. The callout order for a counterbored hole is:
    a. hole size, counterbore size, counterbore depth.
    b. counterbore depth, counterbore size, hole size.
    c. hole size, counterbore depth, counterbore size.
    d. None of the above.

33. A spot facing operation is used to:
    a. provide a bearing surface.
    b. remove a burr.
    c. Neither of the above.
    d. Both (a) and (b).

34. Leaders for local notes that pertain to a hole should extend:
    a. to the center of the hole.
    b. to the edge of the hole.
    c. Neither of the above.
    d. Both (a) and (b).

35. Good dimension practice is based on standards given by the:
    a. American Institute of Industrial Engineers.
    b. American National Standards Institute.
    c. Dimensioning Council of America.
    d. None of the above.

36. Operating personnel should not have to:
    a. add or subtract dimensions.
    b. multiply or divide dimensions.
    c. scale dimensions.
    d. All of the above.

37. When applying dimensions, one should remember:
    a. reliability, cleanliness, completeness, readability.
    b. accuracy, clearness, completeness, readability.
    c. reliability, clearness, nomenclature, readability.
    d. accuracy, cleanliness, nomenclature, readability.

38. The letter R refers to:
    a. radians.
    b. a revision.
    c. the radius of an arc.
    d. the diameter of a circle.

39. Which of the following symbols would specify a spotface?
    a. ⊤.
    b. ⊔.
    c. √.
    d. spf.

40 Notes that do not require a leader are called:
    a. general notes.
    b. local notes.
    c. drawing notes.
    d. machining notes.
    e. None of the above.

41. Which of the following symbols specifies a depth?
    a. √.
    b. S.
    c. ⊤ .
    d. DEPTH.

42. The symbol $\times$ is read as:
   a. times.
   b. by.
   c. Neither of the above.
   d. Both (a) and (b).

EXERCISE 8-1

Name: _____

Fill in the blanks or circle the correct answer.

1. Variation is caused by _____ , _____ , the _____ , and _____ _____ .

2. A stapler would have _____ limits than a watch.

3. The _____ the limits, the more expensive the manufacturing process.

4. _____ are the extreme dimensions of a part feature.

5. Limit dimensioning and plus and minus tolerancing are two forms of _____ dimensional tolerancing.

6. The least material condition of a hole is the _____ hole.

7. The maximum material condition of a shaft is the _____ shaft.

8. Selective assembly matches large holes with _____ shafts and small holes with _____ shafts.

9. The exact theoretical dimension without tolerance is the:
   a. allowance.
   b. maximum material condition.
   c. basic dimension.
   d. datum.

10. The smallest shaft or the largest hole is the:
    a. regardless of feature size.
    b. maximum material condition.
    c. basic size.
    d. least material condition.

11. Which of the following is *not* a basic way of expressing tolerance?
    a. Direct dimensional tolerancing.
    b. Accumulation.
    c. Notes.
    d. Geometric tolerancing.

12. A rectangle with symbols, letters, and/or numbers is the technique for expressing:
    a. basic dimension.
    b. datum.
    c. feature control frame.
    d. All of the above.

13. The rule that states RFS applies with respect to the tolerance, the datum, or both unless MMC is specified.
    a. Rule 1.
    b. Rule 2.
    c. None of the above.

14. Define tolerance. _____

_____

15. Define feature and feature of size. _____

_____

_____

16. Sketch the symbols for the following:

a. Perpendicularity _____   **b.** Total runout _____

c. Cylindricity _____   **d.** Position _____

e. Straightness _____   **f.** Circularity _____

g. Symmetry _____   **h.** Flatness _____

i. Profile of a line _____   **j.** Concentricity _____

k. Circular runout _____   **l.** Profile of a surface _____

m. Parallelism _____   **n.** Angularity _____

17. Sketch the symbols for the following:

a. MMC _____   **b.** RFS _____

c. Diameter _____   **d.** LMC _____

e. 35-mm basic dimension _____   **f.** Datum P _____

g. Between _____   **h.** Statistical tolerance _____

i. A straightness tolerance of 0.2 _____

EXERCISE 8-2

Name: _____

Answer the questions.

1. Indicate the method of specifying the tolerance.

$\overleftrightarrow{\frac{50.1}{49.7}}$

Answer: _____

$\longleftarrow 49.7 \,^{+0.4}_{\ \ 0} \longrightarrow$

Answer: _____

$\longleftarrow 50 \,^{+0.1}_{-0.3}$

Answer: _____

$\phi\, 49.7 -50.1$

Answer: _____

2. Name the three techniques of dimensioning that are associated with tolerance accumulation.

Answer: _____   **Answer:** _____

Answer: _____

3. Determine the tolerance accumulation between surface B and E for the three situations below. Also specify the method.

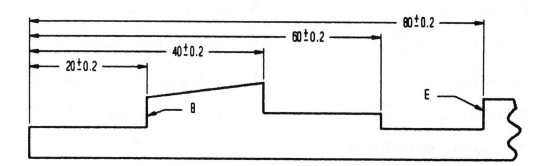

Answer: _____

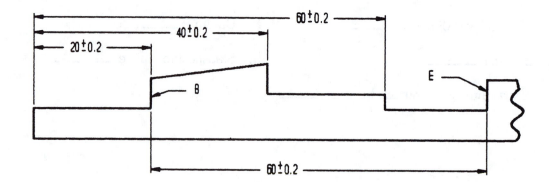

Answer: _____

320

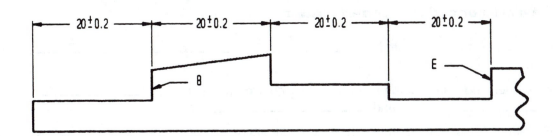

Answer: _____

4. Based on rule 1, determine the variation in form for a shaft with dimensions of ∅ 15.3-15.5. Express your answer in the same way as the text.

5. Based on rule 1, determine the variation in form for a hole with dimensions of ∅ 18.8-19.1. Express your answer in the same way as the text.

EXERCISE 8-3

Name: _____

Fill in the blanks or circle the correct answer.

1. Datum features are associated with _____ or _____ equipment.

2. The sequence of listing datums in the feature control frame is _____ , _____ , and _____ .

3. For cylindrical parts, the datum is represented by two _____ planes that intersect at the axis of the cylinder.

4. The datum symbol for a cylindrical part is placed on _____ .

5. A datum surface does not have a _____ tolerance.

6. A chuck on a lathe is a good example of _____ .

7. Datum targets are designated _____ , _____ , or _____ _____ _____ .

8. Three mutually perpendicular planes are called the:
   a. primary datum feature.
   b. tolerance of perpendicularity.
   c. datum reference frame.
   d. None of the above.

9. The number of points of contact for a primary datum is:
   a. one.
   b. two.
   c. three.
   d. six.

10. The number of points of contact for a secondary datum is:
    a. one.
    b. two.
    c. three.
    d. six.

11. Which of the following are datum features?
    a. surface.
    b. size.
    c. location.
    d. (a) and (b).

12. Where additional support is needed and accuracy is not critical, the datum target is:
    a. a point.
    b. a line.
    c. an area.
    d. (a) and (b).

13. Sketch datum target points for an invisible secondary datum. They are 40 mm from one edge and spaced 60 mm apart. Use the letter F to identify the pins.

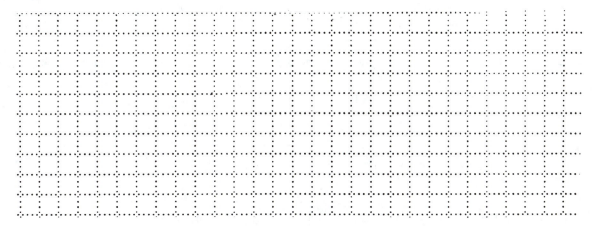

14. Sketch a visible datum target line for a tertiary datum 25 mm from an edge. Use the letter G to identify the pin.

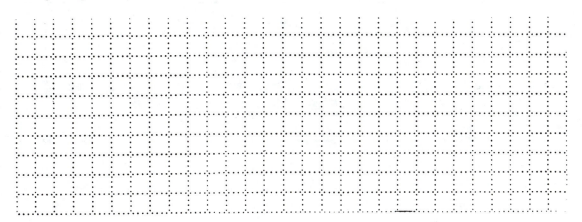

15. Sketch visible target areas (∅ 10 mm) for a primary datum for a cylindrical surface that is ∅ 80 mm. Use the letter H to identify the pins.

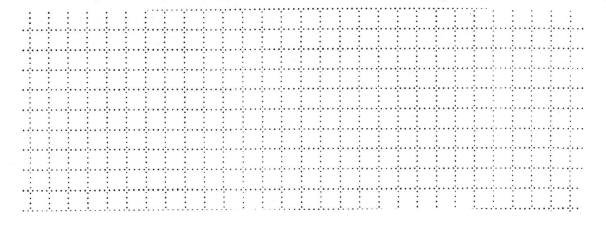

EXERCISE 8-4

Name: _____

1. Sketch the appropriate tolerance *symbol* for each definition:

   _____ is a condition of a surface of revolution where all points of the surface intersected by any plane perpendicular to a common axis or passing through a common center are equidistant from the center.

   _____ is the condition of a surface or axis at a specified angle from a datum plane or axis.

   _____ provides control of circular elements of a surface.

   _____ is the condition in which an element of a surface or an axis is a straight line.

   _____ is a condition of a surface of revolution in which all points of the surface are equidistant from a common axis.

   _____ is the condition of a surface having all elements in one plane.

   _____ is the condition of a surface, center plane, or axis at a right angle to a datum plane or axis.

   _____ is the condition of a surface equidistant at all points from a datum plane or an axis equidistant along its length to a datum axis.

   _____ is the condition where the axis of all cross-sectional elements of a surface of revolution is common to the axis of a datum reference.

   _____ is a condition in which a feature or features are symmetrically disposed about the center plane of a datum feature.

   _____ a zone in which the center, axis, or center plane is permitted to vary from the exact location, which is referred to as its true position.

   _____ provides combination control of all surface elements.

2. Where straightness is specified for an axis, the feature control frame is placed with the
   _____ _____ _____ .

3. Tolerances of form do not require the specification of a _____ .

4. Circularity is a combination of _____ and _____ .

5. Profile tolerancing is unique because it can be applied _____ or
   _____ a datum.

6. Tolerances of orientation require a _____ .

7. Tolerances of location deal with _____ _____
_____ .

8. Tolerances of location require at least one datum and _____ .

9. Square tolerance zones are not recommended because they are difficult to _____ .

10. Where positional tolerance is applied at MMC for a surface, the theoretical boundary is the hole MMC minus the _____ _____ .

11. Concentricity specifications should not be used because _____ is difficult.

12. FIM stands for _____ _____ _____ .

13. Runout is most frequently used on _____ objects.

14. Sketch the meaning of ☐ 0.4

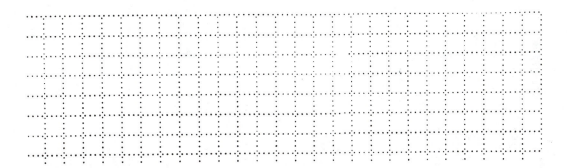

15. Given ⟶ 0.4 for a feature of size (shaft) where Ø 15.3-15.7, determine the virtual condition.

EXERCISE 8-5

Name: _____

1. Distinguish between ⓪ and ⟡ either graphically or verbally.

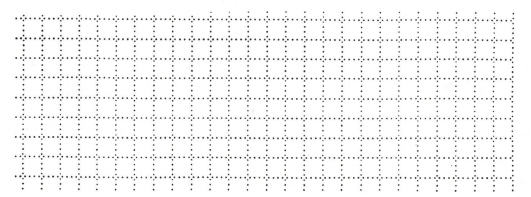

2. Distinguish between a bilateral and a unilateral tolerance for ⌒.

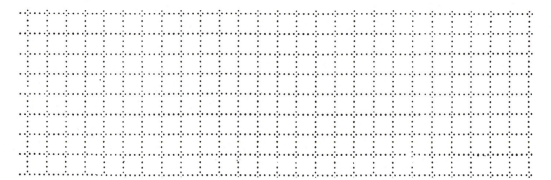

3. Sketch the meaning of ∠ 0.5 D .

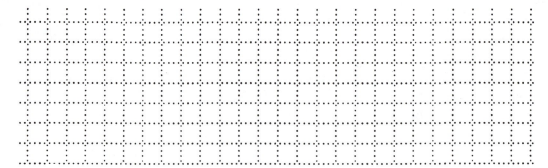

4. Sketch the meaning of // 0.2 J .

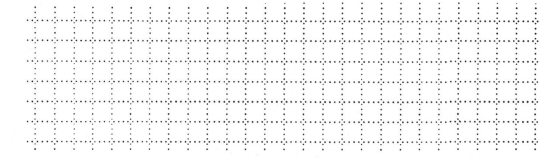

EXERCISE 8-5 (Continued)

Name: _____

5. Given ⟂ | ⌀ 0.3 | B Ⓜ for a feature of size (shaft) of ⌀ 32.4-32.8, determine the tolerance zone at MMC and LMC. What is the bonus tolerance?

6. Sketch a hole (⌀ 20 ± 0.2) and properly dimension its true position with a 0.3 diameter tolerance zone at MMC. The hole is located 35 and 45 mm from a corner that is the intersection of datums A and B.

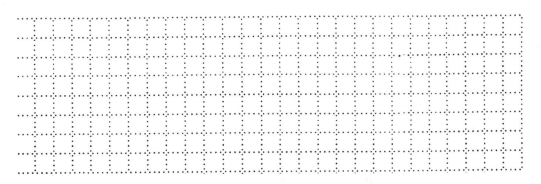

7. Given a hole (⌀ 20.3-20.8) and a positional tolerance of ⌀ 0.4 Ⓜ for the axis, determine the positional tolerance at MMC and LMC. What is the additional tolerance?

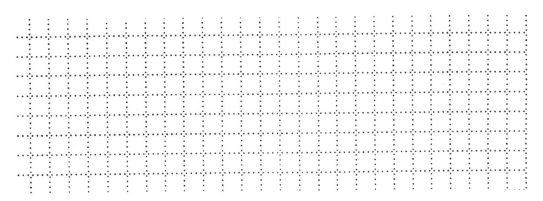

8. Distinguish between circular and total runout by means of a sketch or verbal discussion.

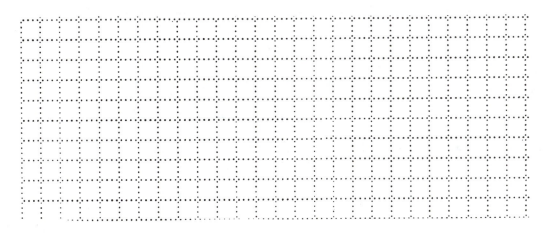

EXERCISE 8-6

Name: _____

Fill in the blanks or circle the correct answer.

1. Information in _____ indicates that it is for reference only.

2. Clearance fits have an _____ _____ between the mating parts.

3. Transition fits sometimes have _____ _____ and sometimes an _____ _____ .

4. The hole basis system uses the _____ _____ as the basic size.

5. The shaft basis system uses the _____ _____ as the basic size.

6. The upper deviation for the shaft in the shaft basis system is _____ .

7. The lower deviation for the hole in the hole basis system is _____ .

8. The fundamental deviation of the shaft is a/an _____ letter and for a hole a/an _____ letter.

9. Define *basic size.* _____

_____

_____

10. Define *upper deviation.* _____

_____

_____

11. Define *lower deviation.* _____

_____

_____

12. Define *fundamental deviation.* _____

_____

_____

EXERCISE 8-6 (Continued)

Name: ______________________

13. Define *international tolerance.* ______________________

____

____

14. Define *tolerance zone.* ______________________

____

____

15. Using a close-running fit for two mating parts and the shaft basis system, determine the hole and shaft tolerance symbols, basic size, and limit dimensions. The design calculation for the diameter is 7 mm. Also calculate the tightest and loosest fit.

16. Using a locational interference fit for two mating parts and the hole basis system, determine the hole and shaft tolerance symbols, basic size, and limit dimensions. The design calculation for the diameter is 12 mm. Also calculate the tightest and loosest fit.

Name: _____

1. Define *roughness*. _____

_____

_____

2. Define *surface texture*. _____

_____

_____

3. Define *roughness height*. _____

_____

_____

4. Define *roughness width*. _____

_____

_____

5. Define *roughness-width cutoff*. _____

_____

_____

6. Define *lay*. _____

_____

_____

7. Define *waviness*. _____

_____

_____

8. Define *waviness height.* _____

_____

_____

9. Define *waviness width.* _____

_____

_____

10. Express a roughness width of 0.1, roughness width cutoff of 0.3, waviness height of 0.008, roughness height of 2.3, parallel lay, and waviness width of 4 in the symbol below.

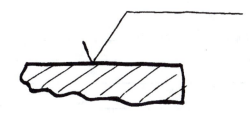

11. Which metal removal processes would be recommended for a roughness average of 0.80 μm?

12. Which metal removal processes would be recommended for a roughness average of 0.20 μm?

EXERCISE 9-1

Name: _____

Fill in the blanks.

1. Screw threads are classed as _____ fasteners.

2. Threaded fasteners are manufactured in standard _____ and _____ .

3. Threads may be either internal or _____ .

4. Thread forms used in U.S. industry are based on approved _____ .

5. Screw threads are used for _____ _____ _____ , _____ _____ , and _____ _____ .

6. Depending on the application, a thread may be either left hand or _____ _____ .

7. Lead is the distance a screw thread advances in _____ _____ measured parallel to the axis.

8. In a double thread, the lead is equal to twice the _____ .

9. Threads may be represented on drawings using one of three conventional methods: _____ , _____ , and _____ .

10. To save drafting time, the _____ convention is used.

11. The unified system has _____ graded pitch series, and the metric system has _____ graded pitch series.

12. In a thread note, UNF stands for _____ _____ _____ .

13. In the unified system, a class 3 fit has a _____ allowance between mating parts and has _____ tolerances than either class 1 or 2 fit.

14. A 12 UNC-2 thread would provide _____ adjustments than a 20 UNF-3 thread.

15. In the metric system, a tolerance grade of 8 is a _____ fit than a lower number.

16. In the metric system, the _____ is specified by a letter, and in addition, capital letters stand for an internal thread and lowercase letter for an external thread.

17. A right-hand thread advances into the hole when turned _____ .

18. In the threaded note .25 20 NC-1, the 20 is the _____ _____ .

19. Finished nuts and bolts have a _____ face.

20 Cap screws join _____ parts together.

EXERCISE 9-1 (Continued)   Name: _____

21. Setscrews are used to prevent _____ between two parts.

22. One example of a _____ fastener is the rivet.

23. Welding processes are classified into three main types: _____ ,
    _____ , and _____ .

24. A jam nut is a type of _____ _____ that is used to prevent a
    nut from loosening on a bolt.

EXERCISE 9-2

Name: _____

Fill in the spaces below.

1. In the space provided, sketch the following screw thread representations:

   a. Simplified external.   **b.** Simplified internal

 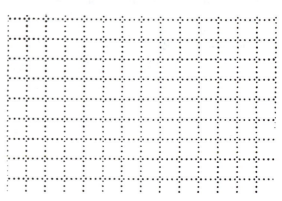

   c. Schematic external   **d.** Schematic internal

 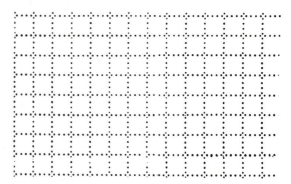

2. Sketch an external thread giving the crest, root, pitch, major diameter, minor diameter, pitch diameter, depth, and angle.

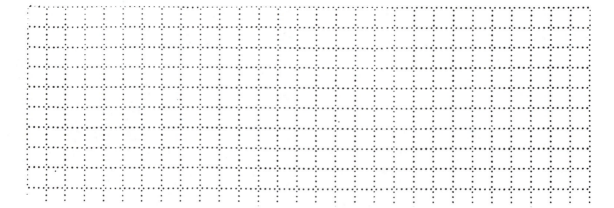

EXERCISE 9-3

Name: _____

1. Using the unified and metric tables, fill in the missing information:

	NOMINAL SIZE (BASIC MAJOR DIAMETER)	PITCH	METRIC/UNIFIED	SERIES
a.	1.6			Constant
b.	4 (0.1120)	1/48 = 0.02		
c.		0.8		Coarse
d.	.4375			UNC
e.	5.5		Metric	

2. Fill in the spaces below.
   a. Interpret the following thread specification: .750-10 UNC-2A.

   .750 _____

   10 _____

   UNC _____

   2 _____

   A _____

   b. Interpret the following specification for an internal thread: .3125-24 UNF-2B-LH.

   .3125 _____

   24 _____

   UNF _____

   2 _____

   B _____

   LH _____

c. Interpret the following specification for a metric thread: M10 × 1.5-6H.

M _____

10 _____

1.5 _____

6 _____

H _____

d. Interpret the following specification for a metric thread: M12 × 1-6g L.

M _____

12 _____

1 _____

6 _____

g _____

L _____

3. Provide the information for the four figures.

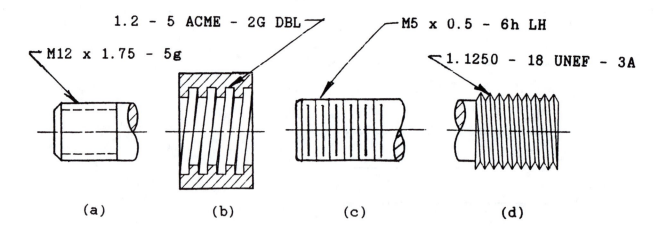

1.2 - 5 ACME - 2G DBL

M5 x 0.5 - 6h LH

M12 x 1.75 - 5g

1.1250 - 18 UNEF - 3A

(a) (b) (c) (d)

	(a)	(b)	(c)	(d)
1. Number of threads per inch.	NA		NA	
2. RH or LH thread				
3. Thread form used				
4. Tolerance (class of fit)				
5. Single or which multiple thread				
6. Pitch of thread				
7. Lead of thread				
8. Major diameter				
9. Internal or external thread				
10. Method of representation				

EXERCISE 10-1

Name: _____

Fill in the blanks.

1. The term "working drawing" includes both _____ and _____ drawings.

2. A _____ drawing is a completely dimensional shop drawing made to scale.

3. A _____ drawing shows one part only.

4. A detail drawing sketch is a completely _____ sketch.

5. A detail drawing and a _____ view are not the same.

6. Reading a drawing involves _____ and _____ .

7. Title blocks provide the _____ _____ about the part or assembly.

8. One reason for a revision to a drawing is _____ _____ .

9. The _____ _____ _____ contains the list of the parts and/or material used on or required by an assembly drawing.

10. The first step in reading a drawing is to analyze _____ _____ _____ .

11. Before a drawing is said to be finished, it must be checked for _____ , _____ , and _____ .

12. The bill of materials shows _____ parts as well as _____ parts.

13. When analyzing the part, concentrate on the _____ view first.

14. The last step in analyzing the part is to study one _____ at a time and begin to picture in your mind the _____ of the real object.

15. A detail drawing illustrates a single component or part, showing a complete and exact description of its _____ , _____ , and _____ .

Answer the questions.

DRAWN: *DAB*	11/13	NORTH AMERICAN UNIVERSITY
CHECK: *GG*	11/15	COLLEGE OF ENGR. AND TECH.
		ANYTOWN, USA
DESIGN:		STEM
APPROVED: *BO'N*	11/20	
APPROVED:		DWG. SIZE A4
MATERIAL:		45438-2
1.25 RD BR ROD		SCALE: 1"=1" WT: .05# SHEET: 1 OF 1

1. What is the name of the part? _____

2. Give the drawing number. _____

3. How many sheets are in the set? _____

4. What is the weight of the part? _____

5. To what scale was the drawing made? _____

6. What material is specified? _____

7. Who made the drawing? When? _____

8. How many people have initialed the drawing? _____

EXERCISE 10-2

Name: _____

Answer the questions.

DRAWN: *Jen*	2/20	NORTH AMERICAN UNIVERSITY
CHECK: *JJ*	2/21	COLLEGE OF ENGR. AND TECH.
DESIGN:		ANYTOWN, USA
APPROVED: *BH*	2/24	VALVE BODY
APPROVED: *BB*	2/25	DWG. SIZE A4 / 45438-1
MATERIAL: FORGING 11894		SCALE: 1 : 2 / WT: 15.3 Kg / SHEET: 1 OF 2

1. Give the name of the part. _____

2. What is the drawing number? _____

3. How many sheets are in the set? _____

4. Give the scale of the drawing. _____

5. What size of drawing was made? _____

6. How many people have initialed, indicating their approval of the drawing, besides the drafter? _____

7. What material is specified? _____

8. Who approved the drawing? When? _____

9. What is the weight of the part? _____

340

EXERCISE 10-2 (Continued)

Name: _____

Answer the questions.

ZONE	REV.	DESCRIPTION	DATE	APPROVED
		REVISIONS		
C5	1	WAS 42.6 DIA.	07/10/98	*B. Boss*
B4	2	WAS M24 X 3 X 50	09/26/98	*B. Boss*
D4	3	FINISH ALL OVER ADDED	11/30/98	*T L*
A2	4	SECTION B-B ADDED	01/03/98	*B. O'H*

1. How many changes are listed? _____

2. What was the change in revision 1? _____

3. In what zone is revision 2 listed? _____

4. In revision 3, what was added? _____

5. What is the date of the first revision? _____

REVISIONS				
ZONE	REV.	DESCRIPTION	DATE	APPROVED
	A	3 X 45° CHAMFER ADDED	06/12/98	*B. Boss*
	B	WAS REAM AFTER DRILLING	09/15/98	*B. Boss*
	C	NOTE ADDED	02/23/98	*B. O'H*

1. How many changes are listed? _____

2. What was the change in revision B? _____

3. What was added in revision C? _____

4. What was the size of the chamfer added? _____

5. What is the date of the last revision? _____

Name: _____

Answer the questions.

7	1	CAP SC.	1112 ST.	M10 X 1.5 X 50
6	2	STEM	STN. STL.	TYPE 316
5	2	HANDLE	BRASS	Ø 12 ROD
4	1	BUSHING	BRONZE	OILITE #1612
3	1	O RING	RUBBER	BUNA N SYN
2	3	WASHER	1020	CHROME PLATE
1	1	BASE	C.I.	
ITEM	QTY.	NAME	MATERIAL	SPECIFICATIONS
BILL OF MATERIALS				

1. What is the block called that lists the materials? _____

2. How many different parts are listed? _____

3. Give:

   a. the name of item 3 _____

   b. the material _____

   c. the specifications _____

4. For the stem give:

   a. the material _____

   b. the specification _____

EXERCISE 10-3 (Continued)

Name: ____________________

5. For item 2 give:

a. the quantity required ____________________

b. the part name ____________________

c. the material ____________________

d. the material finish ____________________

6. For the handle give:

a. the item ____________________

b. the specification ____________________

c. the material ____________________

d. the quantity ____________________

Answer the questions.

5	2	NUT	1112 ST.	M12 X 2 SQ.
4	2	WASHER	1112 ST.	M12
3	1	BUTTON	BRONZE	Ø 12
2	6	SPRING	MUS. WIRE	1 MM
1	1	VALVE	C.I.	PATTERN NUMBER 352-K
ITEM	QTY.	NAME	MATERIAL	SPECIFICATIONS
BILL OF MATERIALS				

1. How many different parts are listed? ____________________

2. Give the name of item 5. ____________________

3. For the valve give:

    a. the specification   _____

    b. the material   _____

4. For the spring give:

    a. the quantity required   _____

    b. the material   _____

    c. the specification   _____

5. What material is required for the button?   _____

6. For item 5, what is:

    a. the part name   _____

    b. the specification   _____

EXERCISE 10-4

Name: _____

Answer the questions.

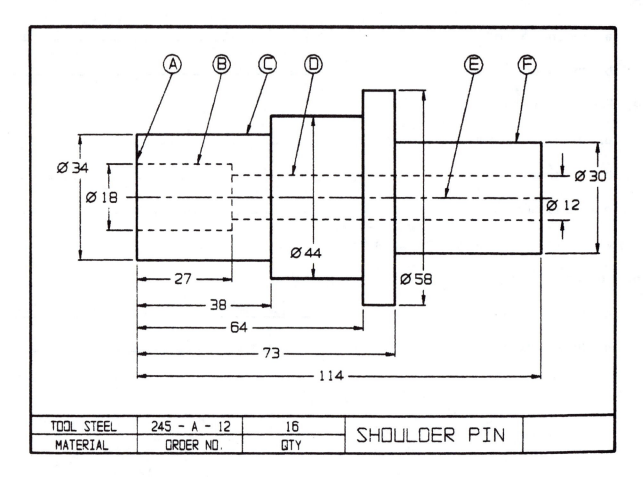

TOOL STEEL	245 – A – 12	16	SHOULDER PIN	
MATERIAL	ORDER NO.	QTY		

1. Name the view represented. _____

2. What is the shape of the shoulder pin? _____

3. How many outside diameters are shown? _____

4. What is the largest diameter? _____

5. What diameter is the smallest hole? _____

6. What is the overall length of the pin? _____

7. How deep is the Ø18 hole? _____

8. How wide is the Ø44 portion? _____

EXERCISE 10-4 (Continued)

Name: _____

9. What letters represent object lines? _____

10. What kinds of lines are **B** and **D?** _____

11. What letter represents the centerline? _____

12. What does the centerline indicate about the holes and outside diameters? _____

13. Give the width of the Ø58 portion. _____

14. State the order number of the part. _____

15. What is the material specified for the pins? _____

Answer the questions.

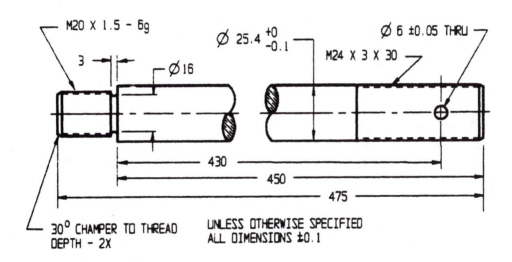

1. Give the finished length, including the tolerance permitted. _____

2. What size of hole is specified? _____

3. Interpret the thread specification on the right end. _____

4. Give the dimensions for the thread relief. _____

347

5. How long is the thread on the left?   _____

6. Give the limit dimensions for the shaft diameter.   _____

_____

7. Why are the break lines used?   _____

8. What is the chamfer specification on the left end?   _____

9. Explain the tolerance and allowance specification on the M20 thread.   _____

Answer the questions.

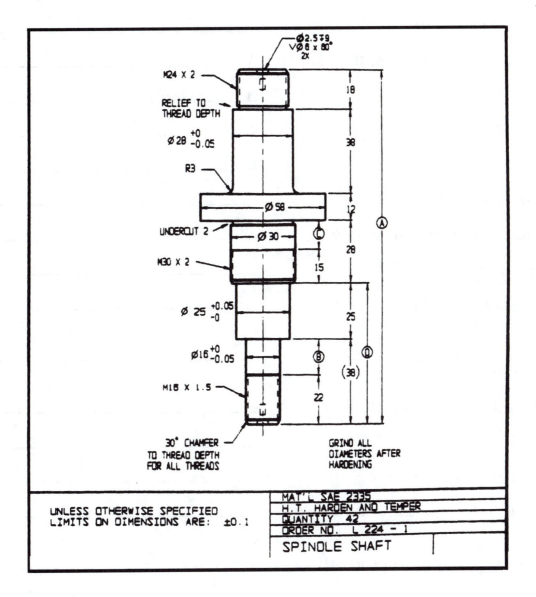

1. What is the name of the part? _____

2. What is the material specification? _____

3. How many parts are required? _____

4. What type of heat treatment is called for? _____

5. What is the order number?   _____

6. What is the hole depth?   _____

7. Find the following lengths: **A, B, C, D.**

    A _____

    B _____

    C _____

    D _____

8. What is the largest diameter?   _____

9. What is the largest thread diameter?   _____

    Smallest thread diameter?   _____

10. What is the tolerance of the ∅28 dimension?   _____

11. What is the length of the ∅25 portion of the shaft?   _____

12. What is the length of the M30 thread?   _____

    The M16 thread?   _____

13. Which thread is a coarse series?   _____

14. What is the final operation on the diameters?   _____

15. What does the (38) dimension mean?   _____

Answer the questions.

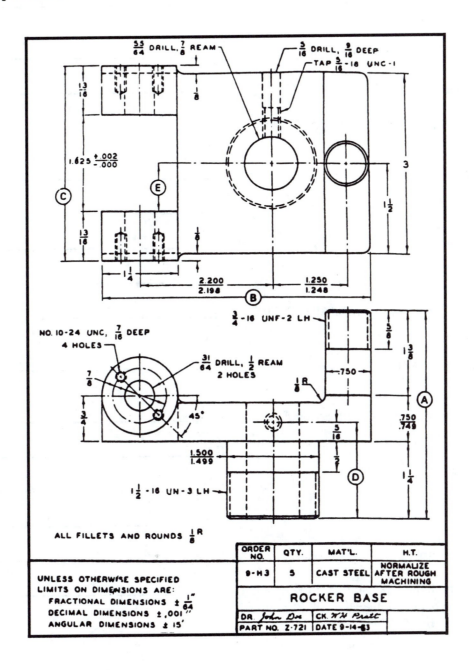

1. What is the name of the part? _____

2. What type of material is used? _____

3. Is this part heat-treated? What type of treatment is specified? _____

4. How many general notes are there? _____

5. How many local notes? _____

6. What are the limits on dimensions for the following:

   a. Fractional _____

   b. Decimal _____

   c. Angular _____

7. How many holes are reamed? _____

8. What is the length of the 1½-16UN-3LH thread? _____

9. What radius is specified for fillets and rounds? _____

10. What are the lengths for the following: **A, B, C, D, E?**

    A _____

    B _____

    C _____

    D _____

    E _____

11. What are the limit dimensions for the center line distance between the $^{55}/_{64}$ drill $^{7}/_{8}$ ream hole and ¾-16UNF-2LH? _____

12. Describe the following thread callout:

1½ _____

16 _____

UN _____

3LH _____

13. Describe the following local note callout:

NO. 10 _____

24 _____

NC _____

⁷⁄₁₆ _____

4 _____

14. Express the centerline distance between the ⁵⁵⁄₆₄ drill, ⅞ ream hole and ³¹⁄₆₄ drill, ½ ream 2 holes as a plus-and-minus tolerance dimension.

_____

15. What is the length of the unthreaded part of the 1½- 16UN-3LH thread?

_____

Answer the questions. See the print on the following page.

1. How many surfaces are to be finished?   _____

2. Except where noted otherwise, what is the   **A** _____
   tolerance on **A** and **B?**
                                               B _____

3. What is the tolerance on the Ø12 holes?   _____

4. What are the limit dimensions for the 40 di-   _____
   mension shown on the side view?

5. What are the limit dimensions for the 25 di-   _____
   mension shown on the side view?

6. What is the maximum permissible distance   _____
   between the centers of the Ø8 holes?

7. Express the Ø12 hole as a plus-and-minus   _____
   tolerance dimension.

8. What is the maximum tolerance placed on   _____
   dimension **H?**

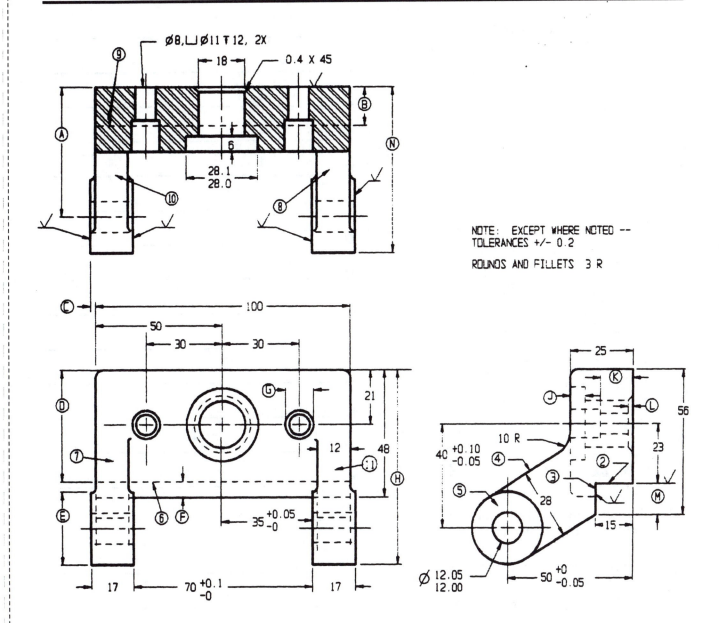

NOTE: EXCEPT WHERE NOTED --
TOLERANCES +/- 0.2

ROUNDS AND FILLETS 3 R

9. Calculate distances **A** to **N:**

    A _____

    B _____

    C _____

    D _____

    E _____

    F _____

    G _____

    H _____

    I _____

    J _____

    K _____

    L _____

    M _____

    N _____

10. What are the limit dimensions for the 6 dimension shown on the top view?

    _____

11. Locate surfaces **4** on the top view.

    _____

12. How many surface **5**'s are there?

    _____

13. Locate line **3** in the top view.

    _____

14. Locate line **6** in the side view.

    _____

15. What surfaces in the front view indicate line **4** in the side view?

    _____

EXERCISE 11-1

Name: _____

Fill in the blanks or circle the correct answer.

1. CAD/CAM is the use of computers to aid the design and _____ _____ of a product.

2. The first step is usually to communicate ideas by means of a _____ _____ .

3. CAD/CAM are _____ activities.

4. CAD/CAM is an outgrowth of the development of _____ .

5. Menus give easy access to specific instructions or drawing _____ .

6. Dot-matrix printers are relatively _____ , but line quality is not as good as a plotter.

7. One major advantage of CAD is that the link with _____ is easily effected.

8. The primary input methods are keyboard, digitizer/puck, mouse, and _____ .

9. Mirroring is used with objects that are _____ .

10. The copy function saves the drafter considerable time when a number of _____ is repeated.

11. CAD is defined as:
    a. computer-aided design.
    b. computer-aided drafting.
    c. computer-aided design and drafting.
    d. All of the above.

12. Three types of magnetic storage media are:
    a. cartridge, floppy disk, hard disk.
    b. cartridge, floppy disk, paper tape.
    c. cartridge, hard disk, paper tape.
    d. hard disk, paper tape, floppy disk.

13. A small, low-cost computer is called a:
    a. maincomputer.
    b. PC.
    c. minicomputer.
    d. small computer.

14. Which of the following is *not* a component of a computer system?
    a. storage.
    b. production.
    c. input.
    d. output.

15. Which of the following is *not* a function of a workstation?
   a. Interface with the display monitor.
   b. Provide digital descriptions of the graphic image.
   c. Facilitate communication between the user and the system.
   d. All of the above.

16. Which of the following is an output device?
   a. keyboard.
   b. digitizer/light pen.
   c. printer.
   d. mouse.

17. The best device for inputting textual material is a:
   a. keyboard.
   b. printer.
   c. mouse.
   d. digitizer/puck.

18. A particular location on the screen is given by a crosshair called a:
   a. pointer.
   b. bracket.
   c. cursor.
   d. cross-hair.

19. Which of the following are commercially available software systems?
   a. CADKEY.
   b. VERSACAD.
   c. AUTOCAD.
   d. All of the above.

20 The zoom function enables the system to:
   a. increase the image speed.
   b. make the image larger.
   c. speed up the plotter.
   d. None of the above.

21. The process of turning a part displayed on the display monitor through a selected angle about an axis is called:
   a. rotating.
   b. turning.
   c. moving.
   d. (a) and (c).

22. Which of the following are common CAD system enhancements?
   a. Crosshatching.
   b. Consistency.
   c. Automatic dimensioning.
   d. (a) and (c).

23. As compared to instrument drawing, which of the following is not an advantage of CAD?
   a. Faster.
   b. More accurate.
   c. More consistent.
   d. Neater.
   e. None of the above.

24. Geometric modeling is also called:
   a. solid modeling.
   b. trigonometric modeling.
   c. calculus modeling.
   d. none of the above.

25. A rendered drawing can help:
    a. a designer visualize a component.
    b. an executive make important company decisions.
    c. a salesman show a product to a potential customer.
    d. all of the above.

EXERCISE 12-1

Name: _____

1. The graphics window contains:
   a. a titlebar.
   b. pull-down menus.
   c. the command line.
   d. all of the above.

2. The most comprehensive set of commands available in AutoCAD LT are:
   a. the toolbox buttons.
   b. the toolbar.
   c. pull-down menus.
   d. the coordinate display window.

3. To complete the line drawing command requires:
   a. clicking the left mouse button.
   b. pressing the enter key on the keyboard.
   c. clicking the right mouse button.
   d. double-clicking the left mouse button.

4. Automatic save can be initiated through:
   a. opening the Preferences dialog box in the File menu.
   b. opening the Settings menu.
   c. opening the Assist menu.
   d. none of the above.

5. The Zoom command is used to:
   a. move quickly from one area to another in the drawing window.
   b. increase the mouse speed.
   c. enlarge an area of the drawing.
   d. control the size of circles and arcs.

6. When starting a new drawing there is:
   a. one line type and one layer available.
   b. one line type and multiple layers available.
   c. multiple line types and one layer available.
   d. multiple line types and multiple layers available.

7. Drawing lines in the Ortho mode:
   a. draws lines on a 15° angle.
   b. draws only horizontal lines.
   c. draws only vertical lines.
   d. draws lines only on a horizontal or vertical axis.

8. The offset command will:
   a. create a mirror image of a selected line.
   b. rotate a line to a position 90° to its original position.
   c. remove small marks left on the screen from mouse moves.
   d. duplicates a line at a specific point on the drawing that you specify.

9. Make an isometric drawing of Figure 12-17.

10. Make detail drawings of the individual parts of a mechanical can opener containing at least 5 pieces. Assume any unknown dimensions.

11. Make detail drawings of the individual parts of a mechanical pencil sharpener containing at least 6 pieces. Assume any unknown dimensions.

12. Make an assembly drawing of a mechanical can opener containing at least 5 pieces.

13. Make a detail drawing of a bench vise.

14. Draw a section view of a bench vise assembly.

15. Make an isometric drawing of a small garden cart.

INDEX